SMART AQUACULTURE

SMART AQUACULTURE

Edited by

Dr. Prashant Telvekar

Delve
PUBLISHING

www.delvepublishing.com

Smart Aquaculture

Dr. Prashant Telvekar

Delve Publishing
4164 Lakeshore Road
Burlington, ON L7L 1A4
Canada
www.delvepublishing.com
Email: orders@arclereducation.com

ISBN: 978-1-77956-405-4 (Hardcover)

ABOUT THE EDITOR

Dr. Prashant A. Telvekar (1979) is presently serving as Assistant professor in Department of Fisheries Extension, Economics and statistics, College of Fisheries Science, Nagpur. He has obtained Bachelor of Fisheries Science (B.F.Sc.) in 2000 from Dr. Balasaheb Sawant Konkan Krishi Vidyapeeth, Dapoli. He has done his Masters in Fisheries Science (MF.F.Sc.) in 2002) from the Central Institute of Fisheries Education, Mumbai (CIFE) and Ph. D. in Fish and Fisheries Science with specialization in Fisheries Resources Management from CIFE, Mumbai in the year 2007. He started his career as Assistant Professor of Fisheries Biology at College of Fishery Science, Nagpur in the year 2006. Since 2007, he has been working as Assistant professor in Fisheries Resources, Economics, Statistics and Extension Education at the College of Fishery Science, Nagpur under the Maharashtra Animal and Fishery Science University, Nagpur. His field of specialization is Fisheries Resources Management and Fisheries Extension. He has published 21 Research Papers in the National and International journals of repute. He has participated in many National, State seminars

TABLE OF CONTENTS

LIST OF FIGURES

LIST OF TABLE

ABSTRACT

Fishery products are crucial in ensuring food security and nutrition strategies at all levels. Fish possess significant potential to serve as a valuable alternative source of protein in lieu of livestock and poultry. The global fishing industry has facing challenges such as scarcity of labor, the deterioration of aquaculture water quality, outdated aquaculture methods, human activities, industrial pollution, and an increased occurrence of diseases in aquatic products. Concurrently, the appropriate areas for aquaculture will gradually decrease, leading to a decline in the profitability per unit of land. Insufficient scientific and technological innovation and engineering technical support exist throughout the entire fishery industry chain. Insufficient fundamental research and theoretical understanding exist regarding the interaction between biological environments and the integration of engineering processes. The progress in aquaculture engineering, mechanization, information technology, and equipment construction is significantly inadequate. The solution to these problems is Smart aquaculture involving intelligence and automation. A smart solution to address these gaps in the aquaculture ecosystem is a blended approach – a combination of human and technological intervention to ensure that farmers are aware of the best practices prevalent in aquaculture. Smart technology is the key to better productivity and disease management. From the internet of things (IoT) and artificial intelligence (AI) to remotely operated vehicles (ROVs), Intelligent fish farms, Climate Smart aquaculture, Nanotechnology applications in aquauclture systems to effective drug delivery for fish health management and in seafood processing, remote satellite imagery, these new technologies are being leveraged by several emerging players to combat the existing challenges in aqua farming and enhance overall production. Modern intelligent technologies hold prospects in aquaculture to reduce labor, enhance aquaculture production, and be friendly to the environment. Machine learning is a subdivision of artificial intelligence (AI) by using trained algorithm models to recognize and learn traits from the data it watches. To date, there are several studies about applications of machine learning for smart aquaculture including measuring size, weight, grading, disease detection, and species classification. Moreover, there are numerous components of smart aquaculture, such as the collection of information through a variety of temperature, dissolved oxygen, humidity, light, and pH sensors to manage the water quality parameters in the aquaculture system; the transmission of the collected data to the control center via communication nodes; the analysis of data and decision-making stored in cloud platforms; the feedback of decision-making to each execution equipment; and the intelligence to operate a system automatically in order to develop aquaculture in a sustainable and efficient manner that is environmentally friendly.

PREFACE

Aquaculture is an essential component of the agricultural economy, playing a significant role in meeting the global demand for seafood. Historically, the industry has faced challenges, with major farming accidents often stemming from reliance on subjective experience and lack of precise data. Numerous factors influence the water quality in aquaculture systems, including temperature, pH levels, dissolved oxygen, and the presence of harmful substances. Ensuring a stable ecological environment with high water quality is crucial for maximizing production efficiency and maintaining the health and quality of the aquatic organisms being farmed.

With the continuous advancement of science and technology, the integration of intelligence and informatization into aquaculture has emerged as a revolutionary trend. Smart aquaculture leverages technologies such as the Internet of Things (IoT), artificial intelligence (AI), and big data analytics to enable real-time monitoring, prediction, warning, and risk control of the physical and chemical parameters of the aquaculture environment. These technologies also facilitate real-time observation of the characteristics and behaviors of fish, providing insights into the health of the ecosystem and allowing for timely interventions to maintain optimal conditions.

This book, Smart Aquaculture, is a comprehensive guide divided into eight chapters, each focusing on a critical aspect of modern aquaculture. Chapter 1 lays the foundation by explaining the basic principles, history, and significance of aquaculture in meeting global food demands. It sets the stage for the subsequent chapters by providing essential background knowledge.

Chapter 2 deals with the various types of aquaculture systems, including traditional, semi-intensive, and intensive methods. It highlights the advantages and challenges associated with each system, emphasizing the need for innovation to improve efficiency and sustainability.

In Chapter 3, the book explores the concept of smart fish farming. It discusses how advanced technologies, such as automation and data analytics, are transforming traditional fish farms into intelligent systems that optimize production and ensure fish health.

Chapter 4 addresses the integration of the Internet of Things (IoT) in aquaculture. This chapter identifies the obstacles faced in implementing IoT solutions and presents potential strategies to overcome these challenges, thereby enhancing monitoring and management practices.

The focus of Chapter 5 examines the impact of climate change on aquaculture and presents adaptive strategies to mitigate its effects. It underscores the importance of developing resilient aquaculture practices to ensure sustainability in a changing climate.

Chapter 6 introduces the innovative use of nanotechnology to improve water quality, disease control, and feed efficiency in aquaculture systems. It provides insights into cutting-edge research and practical applications of nanomaterials.

In Chapter 7, the book explores how artificial intelligence, particularly machine learning and computer vision, is revolutionizing aquaculture. These technologies enable precise monitoring and predictive analytics, leading to smarter decision-making processes.

Finally, Chapter 8, discusses the principles and practices of organic aquaculture. It highlights the benefits of organic methods and the challenges faced in maintaining standards and certification.

This textbook is intended for a diverse audience, including aquaculture practitioners, researchers, and students. It is particularly valuable for industry professionals seeking to incorporate advanced technologies and sustainable practices into their operations. Academics and researchers will find it a useful resource for the latest developments and innovations in the field.

Chapter 1

Introduction to Aquaculture

Contents

INTRODUCTION

Aquaculture is one of the fastest growing industries for food production worldwide, and its products constitute a major source of protein for human consumption (Ahmad et al., 2021). It is defined as 'the farming of aquatic organisms including fish, molluscs, crustaceans and aquatic plants in selected or controlled environments (Bouelet et al., 2018). It is currently one of the fastest growing food producing sectors in the world. To meet the growing demand for fish products, aquaculture is expanding quickly. According to estimates from the Food and Agriculture Organization (2018), it accounted for 46% of all fish production in 2018, with 52% of that value going toward human consumption. Annual growth in capture production was predicted to reach 62% by 2030 (Kibenge, 2016; Freitas et al., 2020). The increase in global population and the decline in capture output both contribute to this demand (Rahman et al., 2019). There is a fall in or close to stasis in the increase of supply from catch fisheries as a result of the aquaculture sector's rapid expansion. It follows that aquaculture will continue to contribute more and more to fish supplies. In the official aquaculture industry, the farm is sizable, expertly run, and employs cutting-edge technology. It is widespread in industrialized nations including the United States, Japan, Norway, and France, which account for around 8.6% of the world's aquaculture production (WHO, 1999). However, informal aquaculture is made up of semi-organized, low-tech, and underdeveloped infrastructure units, particularly in low-income nations with a food deficit. Even the main aquaculture-producing nations, such China, India, the Philippines, Indonesia, and Thailand, which account for roughly 80% of world aquaculture production (FAO, 1995), include members of this category. Furthermore, as labor would be required in all related businesses, it has the potential to provide employment. Aquaculture has several facets and offers a variety of environments, from freshwater to marine, and from low-tech, computerized closed indoor water recirculation systems to basic, still ponds. As aquaculture industry is most often linked with the farming of fish, besides it also includes the cultivation of other plants and animals present in the water. A variety of species, such as fish, crabs, mollusks, seaweed, and others, are frequently chosen for aquaculture. In this business, fish or other aquatic animals are cultured either in freshwater or in marine environments (Ahmed et al., 2021). Grass carp was the primary fish produced for the last eight years having a share percentage of 10.5% i.e 5.7. million tons among the finfish group. However, Silver carp and Nile tilapia came in second and third, with a share percentage of 8.8% (4.8 million tons) and 8.3% (4.5 million tons), respectively. The molluscs production in 2018 saw significant contributions from cupped oysters (29.5% or 5.2 million tons) and Japanese carpet shells (23.6% or 4.1 million tons). Meanwhile, white leg shrimp dominated the crustaceans group, making up 52.9% (5.0 million tons). The increase in overall production, particularly in finfish species, was attributed to China's policy modifications and the influence of local and international consumers and markets on the production value chain, according to the Food and Agriculture Organization (FAO, 2018).

1.1. THE HISTORY OF AQUACULTURE

Numerous events and factors have shaped the history of aquaculture. Yet indeed, aquaculture has bred its roots all across the Asian islands. This chapter defines and examines all aspects of this fascinating branch of science, including research, techniques for rearing fish, feed, and medicine for both fish and humans, together with important commercial enterprises like the Alaska Salmon Canning Company and shrimp farming in Thailand. (Orville & David, 1979) For example, by the 15th century, it was possible to breed fish in ponds in Bologna. By 1568, all over Europe, ponds were being built and stocked with carp or trout exclusively, as they were considered to be food sources fit for animals as well as humans (Nash, 2010) (Figure 1.1).

Figure 1.1. *Ancient art illustrating early forms of fish farming.*

This history is not only associated with the civilized modern man, but it also tells of human capture, habitat modification, and reintegration over a period of approximately 10,000 years. It started with the cultivation of wild pulses in the Neolithic period and continues to this day with modern aquafeeds derived from aquatic organisms. (Agarwal, 2019) The earliest evidence of aquaculture is a 6,000-year-old account of fish domestication in China. The term "aquaculture" is a portmanteau of the terms "aqua" or "water" and "cultivation," which comes from the Latin term "culturare." The global aquaculture industry produces appro□imately 1 billion tons, or roughly 30% of total global fish production, with 0.1% accounting for 80% of market value (Nash, 2010).

1.1.1. Earliest Evidence of Aquaculture

The history of aquaculture, which began about 12,000 years ago, has been the most significant human-related evolution in the evolution of fish, shellfish, and crustaceans. The reductionist idea that environmental and genetic factors could explain behavior was the foundation of aquaculture's success. The earliest evidence of aquaculture in the world has been found at an archaeological site dated to 10,000 BC. More precisely, most sources bear roots in China during the Neolithic period (8500 BC). Fish were kept in ponds, kept in cages, and fed on kitchen scraps. This was sourced from the verses from the Shijing (or "Classic of Poetry") written around the year 1000 BC. It is here that we find references to fish farming, and it was this practice that led to the development of aquaculture in other parts of Asia and Europe.

In 300 BC, the Greek philosopher Aristotle described certain types of fish as being food for humans. These fish were seen to swim in tanks and live on fruit and vegetables while they grew to marketable size. In 3500 BC, an Egyptian prince cultivated oysters to feed his family and guests. In 2600 BC, the Chinese began fish farming using rice paddies as farms. By 2000 BC, farmers in China had begun raising carp in ponds on which they planted rice paddies. In Europe during 3000-2000 BC, freshwater sturgeon was another early form of farmed fish (Habgood, 2013).

Some evidence also dates back to 3400 BC, in the area around modern Khuzestan in southern Iran. According to ancient texts from this region and from Egypt, the first captive fish were kept in pens near the Nile. This practice continued for several centuries before the cultivation of highly domesticated species was mastered and brought about by tsunamis, possibly as a result of human overpopulation in some regions. (Orville & David, 1979).

Most texts also acknowledge the rise of aquaculture during the reign of Caliph Harun al-Rashid. Caliph Harun al-Rashid was recorded as having visited an artificial pond in 786 AD that had been constructed to raise fish and vegetables with no visible support. This period is considered to be the golden age of Arabic science and culture. Freshwater fish were frequently traded along the Mediterranean. The earliest written documentation, which dates back to this period, describes how carp were treated with salt and wool in order to protect them from disease caused by parasites. This method was very successful and depended on partial salting or preservation, although it led to substantial losses due to spoilage. He also directed his vizier to create an artificial lake in

Baghdad. Aquaculture during this period consisted of using ponds and lakes for raising fish rather than simple artificial ponds or natural ponds.

In the Egyptian corridors, the earliest evidence of aquaculture dates back some 7,000 years to the Sumerian city-states developing Mesopotamian agriculture, viticulture, and sedentary water buffalo management. In ancient Egypt, once the Nile was controlled by royal dams and canals across the river, the clearing of land for farming planted it with rice. The Islamic Golden Age saw the birth of some of the most important agricultural practices in history: aquaculture, which played a big part in feeding great masses of people. The Islamic Golden Age is an epoch in the history of Islam, spanning from the middle of the eighth century until the end of the twelfth century, with virtually every aspect of life—politics, society, and culture—being transformed by this period (Nash, 2010).

1.1.2. Aquaculture in the Neolithic Period

The Neolithic Period is a period of time between 10,000 BC and 3000 BC. It was the first time in history that boats traveled on the oceans of the world. During this time, the Neolithic people, who lived between Anatolia and China, domesticated fish for food. (Mark et al., 2019) Due to their ability to farm fish, Neolithic people expanded their populations and brought them closer to other people. This involvement with the art of farming has led to many innovations within the industry, such as techniques for managing water quality and habitat management. The earliest Neolithic aquaculture was developed in the Middle East, and it was focused on harvesting shellfish and aquatic bivalves (Habgood, 2013).

Fish farming as of the Neolithic era greatly came to light at the time of the Sumerian civilization. The cultivation of fish was based on observing that a pond could preserve fish and that trapping fish in nets allowed a large number of them to be caught easily. The domestication of animals had important consequences for aquaculture. On the other hand, not all animals are suitable for species farming, and only those that do not damage their environment can be raised in captivity. Another important development was the introduction of aquacultural technology itself. (Mark et al., 2019) It is also during this period that we understand the development of the Sumerian clay tablet. Some scholars appreciate the fact that the oldest reference to fish in the Old World is a Sumerian clay tablet from about the year 3000 B.C. The Egyptians, who had seaworthy ships by that time, began producing and e□porting fish for food at least as early as 1350 BC. The ancient Greeks were known for their love of fish and seafood dishes. The Neolithic man e□ploited fish by building artificial ponds. Culture and religion have long been linked to aquaculture (Nash, 2010).

1.1.2.1. Aquaculture and the Sumerian civilization.

The Sumerian civilization is widely known for its incredible achievements in science and engineering. However, one may never have considered that their creation of an intricate system to cultivate fish is what really made them so great. In fact, the Sumerian clay tablet with aquaculture information is an important piece of history that may have been lost had it not been rediscovered and translated. The Sumerian clay tablet described the return of fish specimens to the Tigris River. Archaeologists discovered the fish at Ur, an ancient Sumerian city-state, and they were 3,900 BC in date. Aquaculture continued into ancient Egypt, where one was found in a tomb dating back to 3400 BC. The ancient Romans also had their own aqua farms and fisheries, where it was said that "fish represented all of nature's balance, for without them we would be near starvation."

Sumerian clay tablets contain thousands of farming accounts, and one particular e□ample has a listing of "fish" that was set in an area that corresponds to what is today Israel. The details of these records make it quite clear that the population was growing in size, improving their agricultural practices, and breeding fish for consumption. The new data suggest that aquaculture played an important role in Mesopotamian agriculture from as early as 5500 BC onwards." The Sumerian clay tablet also includes a detailed account of how to cultivate plants and animals in ponds, including the use of nets and the knowledge of preventing disease in water buffalo (Habgood, 2013).

The Sumerian civilization also played a big role in the development and recognition of the aquatic ecosystem. The earliest evidence of the cultivation of aquatic plants comes from the Sumerian clay tablets. Among these texts, written between 3200 and 2400 B.C.E., was a list of foods including lettuce, grapevine seeds, and fruit, as well as fish taken from ponds and rivers. The Sumerians, as a community, kept carp in large tanks and fed them live fish. These ancient feeding methods have been passed down through history, where fish farming was adopted in China, Egypt, Rome, and other countries around the world. Today, we have modern aquaculture systems with advanced tools and technology to control water quality and protect against disease.

1.2. HISTORY OF AQUACULTURE: THREE PHASES

The development and expansion of aquaculture have been important for human civilization for millennia. The history of aquaculture can be divided into three phases:

1. **The primitive phase**

The earliest and most widely used method of humankind was aquaculture. In this phase, fish were raised in ponds or small basins to provide food for humans. The availability of fish as a food source and fresh water made it possible. Sometimes it was quite successful: some species can be kept in very close association with humans, and relatively little effort needs to be made to raise them. In these early days, the species used were primarily carp and tilapia (Habgood, 2013).

The first use of an aquatic environment for raising fish occurred in 1959, when a Chinese entrepreneur began growing goldfish in Yantai City. The use of fundamental methods and tools, such as suspended fish cages, covered ponds, and feed additives, distinguished the primitive phase of aquaculture (Nash, 2010).

2. **Artisanal phase**

The artisanal phase was the second in the history of aquaculture. More sophisticated equipment like tanks and cages came next, followed by nets and baskets. During this phase, farmers transitioned from growing cages of two or three fish in small ponds and lakes to larger ponds with larger cages that provided more food and space for their hatcheries. Smaller farms grew into large operations that employed workers to maintain their farms, harvest fish from pens as they matured, feed them, and sell them in markets or directly to customers through a buyer's counter. During this time, fishermen used fish farming to supplement their income by selling fish. The artisanal method necessitated knowledge and skill in hatching eggs, raising young, and catching and transporting adult fish. It started in 1969, when Chinese farmers began commercial fishing by purchasing their own fish stocks. (Martinell, n.d.)(Mark et al., 2019).

3. **Modern phase**

This phase, which began in the 1950s and 1960s, focuses on the development of new aquaculture-specific technologies. Ring culture, high density nets, semi-merged cages, and aerated recirculation systems for oysters are all examples of innovative practices. It also includes the development of new types of aquaculture, such as cage culture and recirculating systems. Phase 3 also saw advancements in antibiotics and protein fortification for animal feed. The most recent advances in aquaculture have been linked to improvements in feeding habits and systems. This method employs hydroponics, an enclosed system that allows fish to be fed indefinitely without having to be removed from their tanks. Previously, fish had to be removed for breeding, but with this method, you simply feed them and allow them to breed in a controlled environment.

Other advantages include the ability to harvest immature fish as needed, which reduces waste and increases profits through more efficient harvesting.

1.3. EARLIEST TECHNIQUES OF REARING FISH

Fish farming practices can be traced back to the Stone Age. Archaeologists have determined the age of a fishing spearhead discovered in a cave in Cheddar, England, to be over 700,000 years old. During the Middle Ages, the prevalence of fish farming in Europe and Asia increased as a result of agricultural methods pioneered by ancient civilizations and the contributions of early scientists such as Aristotle. Subsequently, the methods of fish farming became significantly simplified: fishermen employed nets to capture fish, subsequently seizing and preparing them immediately. As individuals acquired knowledge on shrimp cultivation in artificial ponds, they subsequently engaged in shrimp farming.

It is worth noting that although fish farming has been practiced since ancient civilizations, it was not a major focus at that time. The earliest tamed animals that humans have used for food are fish. Throughout history, humans have employed them for a myriad of purposes, encompassing navigation, watercraft, recreation, food, and medicine. Fish cultivation in China can be traced back to 4000 BCE, according to the earliest known records. Subsequently, these techniques were disseminated globally. Aquaculture currently plays a crucial role in providing sustenance to billions of individuals worldwide on a daily basis (Nash, 2010).

The huchen, a freshwater species indigenous to China, was the initial fish to undergo domestication. The huchen was captured in traps due to its nocturnal nature, which made it challenging to capture. The Ch'in and Han dynasties of China were known to have ponds in their capital cities that were stocked with fish. The Neolithic period provides the earliest known evidence of fishing, although it is probable that the practice existed even earlier. Around 2,000 years ago, inhabitants of islands located on the northern coast of present-day Sweden initiated the practice of cultivating fish in ponds along their shores. The initial recorded occurrence of aquaculture in its present form took place when a group of Chinese monks made the decision to attempt the domestication of carp. At night, they confined the fish in enclosures concealed by rice husks, providing them with small fish as sustenance during the day until they reached a suitable size for consumption. These fish are now commonly referred to as goldfish.

The first documented evidence of this form of aquaculture dates back to the 6th century BC in Shandong Province, China. The process of domestication involved purposefully choosing individuals with favorable characteristics, such as color and size. In 1949, the practice of cultivating white and red carp, also known as "four-color" fish farming, emerged as a result of this. The proliferation of carp farming extended across China, yet it garnered momentum at a later stage in Taiwan and Japan. During the Warring States period, it was a frequent occurrence to observe children engaging in the activity of playing with fish in ponds. A crucial determinant of this achievement was the carp's capacity to endure frigid temperatures, implying that this species has been domesticated for an extensive duration. Ancient Egypt has provided evidence of fish that were domesticated, and in Caral, Peru, an excavation uncovered a collection of American cichlids resembling a small zoo (Figure 1.2).

1. The dugout canoes

Figure 1.2. *Rearing fish in dugout canoes.*

The history of aquaculture began around 6500 years ago, when fishermen in the Tigris and Euphrates River valleys began to farm fish in dugout canoes. These early fishermen observed that many species of fish e□hibited a marked preference for one particular kind of habitat—so, for instance, they caught more catfish by fishing in reed beds than near riffles, or more shad by fishing just below dam constrictions than farther upriver. The dugout canoe was developed by the North American Indians, who used this large boat for traveling, hunting, and fishing. Dugouts were traditionally made from hollowed trunks of trees, but later dugouts were made from a single piece of wood cut down to shape and hollowed out by fire (Nash, 2010).

The dugout canoes were designed to handle the swift currents and shallow waters frequent in river deltas. These canoes were developed from the bark fishing boats of the Bayabas tribes, who lived along the banks of the Parana River. The simple frame of their boats was constructed from single straight logs, normally 35 inches (89 cm) long, rounded on both ends with a wide opening through which they entered their cargoes. The dugout canoes were probably used as early as 8000 BC for migrations and fishing, including locations in the Mediterranean. This method of fishing was probably one of the earliest forms of aquaculture used in mainland Europe.

2. The rice paddy culture

Rice paddy culture has been the earliest technique for rearing fish. It is also one of the most efficient and economical because it does not need any special structures or equipment but only some space for growing rice in sloping fields. It was defined as a traditional system where fish are reared in floating cages, often made from bamboo stakes, in shallow water. These cages can be covered with soil, manure, and straw to provide protection from predators such as crabs and barnacles (Habgood, 2013).

This culture method has been used to cultivate many different species of fish, amphibians, and invertebrates, including carp (Cyprinidae), tilapia and roaches (Rutilidae), and tilapia and grass carps (Cichlidae). In the first century AD, China and Japan used it to produce rice, but it quickly spread outside of these two societies. Aquaculture was carried out along rivers and lakes; however, it was only later when farmers began sinking wooden structures into the water that they were able to take advantage of tidal movements to control salinity levels with seawater pumps (Figure 1.3).

Figure 1.3. *Rice paddy culture, the earliest technique of rearing fish.*

In other terminologies, the method involved growing a particular crop of rice, which provides food for humans and other animals. Fish were raised in small pools or ponds built near these fields as a source of protein for both humans and animals, as well as fertilizer for the soil. Many may also argue that the Chinese had earlier invented a system of controlled flooding that allowed them to grow fish in wetland areas. Eventually, this system was transported to other parts of Asia and became widely practiced in countries across the world (Nash, 2010).

3. The pond culture

The pond culture (also known as bottom culture or simply pond) of fish is one of the earliest forms of aquaculture. It involves the rearing of fish in a simple system with no need for specialized equipment. This method is suitable for small-scale production, home hobbyists, or small farmers who have access to small ponds with only a few fish that they would like to raise. It involves a protected environment where water temperature, light, and oxygen levels can be carefully controlled over time to produce healthy, high-quality fish. The earliest form of this method was used in ponds created by gathering earth or mud from shallow lakes and ponds, then building dams around the pond. Later on, more sophisticated artesian wells were developed that could hold enough water to keep larger populations alive and make it easier to manage larger groups of animals (Figure 1.4).

Figure 1.4. *A pond culture system in the rural town of Masi Manimba.*

The Greeks developed the pond culture, which involved placing fish in a tank of water and letting them grow there. The method was slow, but it worked well at first because there were no other options available. The earliest known example appears to be a bronze statuette from about 1200 BC, which depicts a fisherman who has caught a fish with one hand and is lifting it out of the water with the other hand. An early instance was found in the Hellenistic Kingdom at Perdiccas II's palace in Amyzon, and then Alexander the Great used this technique as well when developing his fish farms (Habgood, 2013).

The first use of fish in Greek culture is described in the Bible. The story of Ceyx and Alcyone relates that their son, Nisus, the king of Megara, had a daughter named Orithyia who was so beautiful that Zeus fell in love with her and made her into a swallow. The trouble started when Apollo chased off this "friend of mankind" by turning her into a partridge. But he also made his sister Artemis' dog do the chasing off job. Artemis then took pity on what remained of Orithyia's human body and changed it into a fish, which she placed in Nisus' well-stocked ponds near Megara. This technology can further be traced back to the people of Sybaris. They were able to create ponds with underground chambers and baskets for the water, which were aerated by the constant movement of water in them. While these techniques cannot be directly linked to the earliest Greek aquaculturists, they were most likely inspired by them.

4. Live culture techniques

Live-culture techniques involve raising fish in a contained tank and feeding them live foods. The growing fish is known as a "cultured" organism because the fish are constantly fed with culture media that are selected for specific traits or characteristics. The earliest techniques developed for rearing fish in aquaculture were those using live-culture techniques. This involves the use of animals to grow cultures in controlled settings. Live culture techniques have been used to rear fish in ponds and tanks since ancient times, but the process was greatly developed after World War II. This development occurred in response to changing attitudes towards food production as well as social and technological developments.

Live-culture techniques are a relatively new development in aquaculture. They allow the producer to control the environment and living conditions of aquatic animals while they are still in their natural habitat. The earliest practice was directly from nature, making use of natural materials found in the initial stage of their life cycle. They include the collection, stabilization, and transfer of juvenile fish, which often present difficulties for those who wish to raise them (Nash, 2010).

5. The submersible pumps

The submersible pumps, known as deep-water paddlesuit pumps, were developed in the early 1900s by a British engineer named William Coggan. These pumps could raise water from depths up to 60 feet, but they required a large power source and couldn't be used easily on land. The submersible pumps were used first in Great Britain and France. The main advantage of this method was the limited number of fish required to be stocked into ponds or tanks, as well as the fact that fewer fish were needed for the same size of production (Habgood, 2013).

6. Polyculture

Originally developed by the ancient Romans, polyculture is the practice of cultivating multiple plants in a single space. It was used to help reduce disease and increase yields, allowing farmers to maximize their crop output. This practice is still practiced today, with fish farming being one of the oldest forms of aquaculture. These farms were initially used to provide food for hunting and fishing e□peditions, but later, these techniques were used to cultivate crops.

1.4. HABITAT MODIFICATION AND RE-INTEGRATION IN EARLY AQUACULTURE

During the development of early aquaculture, habitat modification and re-integration were the primary methods for the creation of sustainable habitats for aquaculture. Early aquaculture techniques were based on capturing wild populations and establishing them in a controlled environment. The most common method used to capture fish was using nets and traps, where adult fish would be captured and released back into the wild. Two other techniques were rearing fish in ponds or utilizing flow-through or recirculating systems (FRA). Habitat modification or re-integration may include changing temperature and salinity, modifying sunlight availability and water depth, adding materials to the environment (sand or gravel), or introducing plants.

However, two methods were highly used in early aquaculture. The first was the invention of ponds and aqueducts to link landlocked lakes with other bodies of water. The second method was the construction of artificial islands, which allowed for more efficient cultivation in small freshwater areas. The early farmers exploited natural reservoirs, marshes, and creeks in an effort to maximize their productivity by exploiting every aspect of their environment. This can also be seen as a natural adaptation because when humans settled down, there was still plenty of greenery around them from areas that had not yet been fully exploited by other people or animals (Nash, 2010).

1.5. AQUACULTURE AND COLONIAL EXPANSION

Like the earliest humans, colonial expansion was the ultimate purpose behind aquaculture. As early as 500 BC, Eurasian colonists began to export fish across

Europe. The growing demand for fish products during this period produced an increase in aquaculture in many other parts of Asia and Africa as well. In coastal areas, aquaculture has been used to support the growing demand for fish. The earliest techniques of rearing fish are believed to have been developed by early man in many locations around the world, including China and Egypt. The history of aquaculture can be traced back to the Egyptians and other civilizations that began fishing for food around 4000 BC. By 1200 BC, the Romans were farming fish in tanks, later called pond culture. Colonial expansion led to new situations where people needed fish more than ever before, including the development of techniques like recirculation systems (using water pumps to move water between pens) and feed storage systems. Aquaculture was one of the agricultural practices that came into use in conjunction with colonial expansion in the late 19th and early 20th centuries. This resulted in the rapid e□pansion of fish farming as an industry in Italy and the Caribbean, most notably Habgood (2013).

1.5.1. Aquaculture and Slave Labor in Colonial Aquaculture

In colonial societies, aquaculture played an important role as a means of dealing with the high demand for food and the provision of raw materials. The supply was required by the colonies, but it also provided slaves for plantations. In addition to the African slave trade, European countries also relied on native labor to help them establish and manage their colonial aquaculture. While it is an income-generating sector, the agricultural sector, however, has been particularly susceptible to labor exploitation by colonial powers due to its reliance on slave labor from Africa and Asia.

During the colonial era of British colonialism in Africa and Southeast Asia, aquaculture became a vital source of income for plantation owners. With this economic boost, slave labor was required to help run the farms and provide for their families. These farmers were forced into slave-like working conditions on their plantations or in fishponds, where they endured long hours of hard labor with little pay or food. The existence of slavery forced Africans away from their traditional ways of food production, which involved long periods of labor in the water and open fields. Slaves were used as manual workers and caretakers for the farms, and they were often punished harshly if they failed to meet their quotas. These harsh conditions made it e□tremely difficult for slaves to survive in the colonies, causing many deaths (Nash, 2010).

1.5.2. Aquaculture and the Spread of Disease in Colonial Aquaculture

The history of aquaculture includes the spread of disease, particularly those diseases that have impacted humans and animals alike. The spread of diseases can be attributed to several factors, including migration patterns, inadequate sanitation and food safety, poor hygiene, and a lack of appropriate infrastructure and knowledge about disease prevention. The location and style of the colonization process, the environment, and financial incentives were just a few of the variables that affected disease spread in colonial aquaculture. The rate of microorganisms that could infect humans was high; this level of infection was so prevalent that mortality was unavoidable even with proper hygiene practices (Habgood, 2013).

In order to establish a successful colony and avoid the spread of disease, aquaculture was an essential part of colonial expansion. Once taken over by European interests, native populations were forced to change their way of life. An outbreak of disease could affect their entire society's economy and potentially cause harm to all involved: the Europeans and the Native peoples alike. Disease spreads through contact with native populations, imported species, and diseases carried by humans to new places. Diseases such as rickettsialpox and malaria were among the most common killers at first, although none of them fully replaced epizootics.

1.5.3. Aquaculture and the Impact of Colonialism on Local Aquaculture

During the colonial era, local communities were often forced to give up their fisheries in favor of aquaculture. The impact that this had on the local economy was severe and contributed to the growth of hunger and poverty. Regardless of whether indigenous people or Europeans directly established the colonies, they frequently had to rely on the local fisheries for food. Colonial and local governments also used aquaculture as a way to increase their commercial power over the native population. This policy can have far reaching effects on indigenous cultures as well as global climate change, which led to a decline in fish populations due to overfishing.

1.5.4. Aquaculture and its Role in the Growth of Colonial Economies

Aquaculture has played an integral role in shaping colonial economies. In the 1600s, European nations began to use aquaculture to establish their own colonial empires. With this growth of agribusiness came a great demand for fish, resulting in movements such as the rise of global aquaculture. Aquaculture helped to stabilize food supplies, provide fresh food, and expand markets for these countries as they moved westward in the 18th century (Nash, 2010). Because Native Americans had little to no knowledge about agriculture, they were forced to sell their food supplies to Europeans. It was important for them to develop methods of cultivation so that they could produce enough food for themselves and others. As a result, some Native American groups began farming fish as opposed to hunting them. In addition to providing food for their tribes, this practice also provided the opportunity for cross-tribal interaction and trade between Native Americans and Europeans.

The earliest known use of aquaculture dates back to China during the third century BCE, when Emperor Shen Nung raised oysters in ponds on farms he founded in central Vietnam. This development also marked the beginning of an industry that would ultimately transport crops throughout Asia via overland trade routes (Habgood, 2013).

1.6. AQUACULTURE AND THE IMPACT OF INDUSTRIALIZATION

As the world has become more industrialized, aquaculture has become a necessity because of its ability to produce food and support human population growth. The industry was pioneered by ancient cultures that used fish for food and medicine, but aquaculture, as we know it today, could not have existed without the advances in industrialized farming and transportation systems. Even though fish farming is a low-impact technique, there are many potential impacts of aquaculture. The primary impact is that some species are at risk of becoming extinct due to large-scale commercial farming and overharvesting. Other impacts include pollution, habitat loss and degradation, water pollution, and changing water quality around the world from chemical residues from fertilizer use and increased acidity from overfertilization.

The history of aquaculture begins with the beginning of mankind, which tried to utilize all available resources. As time went on, this changed into a careful consideration of the environment, safety, and health of the animals. Since then, it has become a multibillion dollar industry that plays a vital role in many aspects of our lives, including food production, biotechnology, biofuel production, and waste recycling. Today, we are able to manage these factors much more efficiently than in earlier times because we have access to modern technology and materials. The history of aquaculture is also closely tied to technological advancements in transportation, refrigeration, and processing (Nash, 2010).

1.6.1. Aquaculture and the Impact of Industrial Pollution

The negative impact of industrial pollution on aquatic ecosystems is far reaching and threatens all of us. As a result, the history of aquaculture has played a major role in educating policymakers and businesses about the need to reduce pollution and protect the environment. However, during the development of aquaculture, human impacts have been significant, including the introduction of disease and pollution through effluent from fish processing plants and industrial runoff into coastal waters. The impact of these factors has been reflected in changes to fisheries worldwide (Habgood, 2013).

Aquaculture can play a role in addressing this problem; however, there are substantial risks associated with using industrial compounds for feed additives, which may result in human health effects and ecological impacts on the marine environment. The history of aquaculture tells us that experimentation began much earlier than many imagine. The history of aquaculture was shaped by both economic forces (many important innovations were driven by rational concerns like profit) and natural selection over time.

Water quality was a major concern in the early days of aquaculture. Pollution from sludge and other waste products caused major fish diseases, which were often fatal to farmed fish. As a result, the first large-scale water quality programs started during the 1950s using filtration media such as activated carbon and sulfate ion exchange (SIX). This technology was still being developed at the time that other forms of pollution began to become important problems for aquaculture: algal blooms and nutrients runoff from agricultural fields bordering on marine habitats for fertilization, as well as excess precipitation runoff from industrial storm water drainage systems entering estuaries and bays where vast expanses of open water exist between land areas and sea walls or other barriers.

1.6.2. Aquaculture and the Decline of Wild Fisheries

One of the greatest impacts on the human diet in recent centuries has been the decline in wild fisheries. Although aquaculture has contributed to this decline, it is difficult to separate the two causes. Aquaculture has been expanding rapidly over the last 30 years and represents a tremendous opportunity for adding protein consumption that can be consumed by all populations (from sea bass in China to salmon in Scandinavia) that are experiencing high rates of growth. The decline in wild fisheries has also greatly contributed to the growth of aquaculture.

First, it allows for the continued sustainability of coastal communities by providing a source of income that can be used to support local economies. Second and most importantly, aquaculture addresses some of the major problems with current fisheries management, such as overfishing, fish stocks becoming more depleted from overuse and pollution, species purity concerns (when wild spawns are introduced into fisheries), overfishing of other species that may be important for future fisheries development, restricted access due to comple□regulations, and inefficient management due to a lack of knowledge about sustainable catch levels or habitats (Nash, 2010).

1.6.3. Aquaculture and the Emergence of New Aquaculture Species

With the advent of industrial agriculture, aquaculture became a powerful tool to help food production in the United States. Although this shift has led to many benefits, it has also introduced new species and cultivars into aquaculture that have had dramatic impacts on their environment. Clearly enough, the history of aquaculture is one of adaptation and evolution. By observing the biological characteristics of different species and the environmental changes that have occurred, it is possible to gain a better understanding of how aquaculture came to develop in the first place (Habgood, 2013).

1.7. AQUACULTURE AND THE RISE OF GLOBAL FISHERIES

In the last century, aquaculture has become an increasingly important source of food for both people and animals. It can be seen as a result of changes in food production worldwide. The use of aquaculture has expanded from producing food for specific human needs to meeting meat needs. Unlike traditional fisheries, aquaculture involves raising aquatic organisms in captivity in order to produce food or other products. The oldest commercial enterprises include;

1. **Alaska Salmon Canning Company**

The first salmon cannery was opened in 1890 by John Joseph Holladay, who had operated a steamboat on the lower Yukon River. Holladay sold his first salmon canned at $2.50 per hundredweight (225 pounds) but found it difficult to compete with lower-priced canned salmon from Chile, so he began making cans of South Sea salmon instead. He soon switched back to Alaska salmon production, however, when prices became high again due to overproduction (Nash, 2010).

The earliest record of commercial salmon canning in Alaska was with the Alaska Salmon Canning Company. In the early 1900s, salmon canning was introduced in Alaska. Prior to this date, most seafood was sold at markets. The first businesses that emerged from Alaska were fish dealers and buying agents. These two were followed by steamboat interests, which later gave way to canneries and processors. First formed in 1909, this company was founded by John Fessenden and Alfred Russell Brown. They built a modern cannery at Cape Krusenstern, where Sitka Sound meets the Bering Strait. The plant produced about 6 million cans per year on an acre of ground; it reflected the highest technical standards of its time.

2. Shrimp farming in Thailand

Shrimp farming in Thailand is one of the earliest commercial enterprises in aquaculture. It has notable significance for the conservation of overall health, natural resources, and food security. Shrimp farming in Thailand started in the 1950s. The industry has grown significantly since then, with 150,000 farmed shrimp ponds now producing between 210,000 and 250,000 metric tons of fish annually. Shrimp farming in Thailand has developed from small-scale family enterprises to large business concerns. The leading producer of farmed shrimp today is Thai Union Shrimp Farming Co., Ltd. Its farms produce about 90% of the world's shrimp exports and are located along both sides of the Mekong River in northern Thailand (Habgood, 2013).

The earliest commercial shrimp farms in Thailand were established in 1988. The industry has since flourished, with a total of 791 shrimp farms and 472 processors over an area of approximately 35,000 ha operated by about 1,000 farmers. The main production areas for farmed shrimp are located along the Gulf coast in southern Thailand, including the coastal waters off Krabi and Phuket Island, as well as those around Ko Samui and Ko Tarutao.

1.8. HUMAN CAPTURE IN THE EVOLUTION OF AQUACULTURE

The evolution of aquaculture has been defined as a series of steps that began with the capture of fish, followed by the selection and breeding of certain traits to increase production. The domestication process was more complex as humans selected traits that allowed them to take advantage of their environment and create new resources. It was a redirection of human energy towards fish breeding, which led to the development of specialized techniques and equipment such as net pens and cages. Indeed, aquaculture is a complex cultural system that has evolved in response to specific environmental factors and human objectives. Aquaculture, as practiced today, is a relatively recent development in human history. Prior to its emergence as an industry, it was practiced by hunter-gatherers, pastoralists, or both. Fishing with nets, traps, and snares became widespread in the early Neolithic period (Nash, C. 2010).

1. The emergence of modern aquacultural practices

The emergence of modern aquaculture practices has transformed the way fish are produced and consumed. In the 19th century, European settlers introduced domesticated salmon and trout to North America. Today, salmon are farmed in tanks up mountainsides from Alaska to Chile. Farmers raise tropical shrimp in ocean pens offshore from coral reefs. Modern aquaculture developed in response to two major developments. The first was the development of antibiotics and vaccines to fight disease, which enabled farmers to grow fish in large pens that were free of infection. The second was the creation of mechanical harvesting methods (such as shoot and gutting) that allowed farmers to maximize their production efficiency at a low cost. The modern era began with the emergence of high-density farming in the 1940s. This invention allowed fish farmers to produce more food with less water, which resulted in cheaper farm prices and increased demand for seafood. Since then, other changes have taken place that have helped make aquaculture one of today's most important agricultural industries globally (Habgood, 2013).

2. Environmental challenges that came with the 21st century

The 21st century has brought a number of environmental challenges that the world has never experienced before. These challenges are grouped into six categories: global changes and climate change; degradation of resources and water; food production; energy and pollution issues; and land use change. The ocean has been acting as a buffer against some of these problems, but that is changing. We need to protect and preserve it so that our children and grandchildren can have secure and healthy oceans in their time as well as our own. According to a World Bank report, aquaculture has significant potential to meet this need by increasing supply while reducing impacts on the environment. With 6,400 species in worldwide commercial production and hundreds more being actively sought out by scientists worldwide, aquaculture has become synonymous with "freshwater farming." Despite the enormous potential value of freshwater fisheries, water pollution from agriculture or industry remains a critical problem throughout much of the world (Nash, 2010).

1.9. OBJECTIVES OF AQUACULTURE

1. Meeting the world's demand

According to the Environmental Defense Fund, global demand for seafood has increased significantly since the 1980s. Given that older adults consume the most seafood of any demographic, an aging population will increase the demand for seafood supplies. The EDF recognizes that fish farming is the only practical way to meet the world's rising demand for fish. Only China and Japan have higher per capita seafood consumption than the United States (Meade, J. W. 2012).

2. Protecting Species

Fish farms typically raise the following species of fish:

- Cod
- Salmon
- Carp
- Tilapia
- Catfish
- European sea bass

These fish might be overfished if not for the safe environments of fish farms, given the increasingly effective fishing methods available to the average angler and commercial fisherman. Aquaculture experts are constantly looking for fish species that can benefit from fish farming while also being protected from extinction.

3. Provides an economic boost

Many state governments have attempted to soften the blow of fishing restrictions in states with thriving fisheries by offering commercial fishermen aquaculture training programs. Coastal communities, in particular, may benefit from having fish farm operations in their communities due to the jobs involved in fish farm construction and operation.

4. Improving quality

Although ongoing improvements in aquaculture can aid in improving filtration, feeding, reproduction, net harvest, and other aspects of fish farming, the safety and quality of seafood can also be enhanced as fish are studied in the controlled environment of fish farms. Researchers can assess the fish's health, eating habits, and reproduction. Though many questions about potential pollution effects in areas near fish farms remain unanswered, scientists and engineers are working to mitigate any adverse effects (Bhujel, 2011).

Other objectives include:

- Development of low-cost, nutrient-dense, palatable, and easily digestible protein-rich human food
- Adding new species and bolstering the populations of fish already present in natural and man-made water bodies through artificial recruitment
- Making ornamental fish for decorative purposes
- Efficient utilization of land and water resources
- Organic waste recycling from humans and animals
- By providing income from commercial and industrial aquaculture
- Sportfish production and recreational fishing assistance

- Development of baitfish for commercial and recreational fishing

Aquaculture uses three types of water: fresh, salt, and brackish. Freshwater, typically abundant in a country's inland regions, and salt water, which is present in seas and oceans, have very different salinities, with fresh water having a salinity of nil and salt water having a salinity of nearly 35 ppt. The salt content of fresh and seawater has a very selective influence on the fauna and flora that inhabit each type of water. The typical inhabitants of each type of water are said to be *stenohaline*, which means they can only withstand a narrow range of salinities in their surroundings. This applies to both finfish and shellfish. *A stenohaline* saltwater fish is a sardine or a mackerel, whereas a *stenohaline* freshwater fish is a carp. Breezy water is found naturally in estuaries, river deltas, lagoons, and backwaters, all of which are subject to tidal regimes. These habitats e□perience significant salinity fluctuations ranging from 0 to 35 ppt, depending on the tide phase and the amount of freshwater discharged through the river into the sea.

Fish and shellfish that live in brackish waters are always *euryhaline*, which means they can withstand significant changes in salinity. *Euryhaline* fish include mullet *(Mugil cephalus*) and mudskipper (*Periophthalmus*), while *euryhaline* crustaceans include several species of *penaeids* (including *Penaeus monodon*) and crabs (e.g., *Scylla serrata*). Finfish and shellfish have different life stages in the ocean, estuaries, and freshwater streams. These animals are either *anadromous* or *katadromous*. *Anadromous* fish naturally breed in freshwater streams but spend most of their lives in the oceans, such as salmon and shad. *Katadromous* fish, unlike eels, migrate from the river to the sea to reproduce. Some types are restricted from moving between freshwater river sections and the estuary. Among the shellfish that go through this life cycle are several species of *palaeomonid* prawns (*Macrobrachium rosenbergi* and *M. vollenhovenii*). Despite spending most of their lives in freshwater, these organisms breed in estuaries. Then some organisms switch between an estuary, a lagoon, and the sea at various stages of their lives. Mullets (e.g., *Mugil cephalus*) and shrimp are finfish and shellfish that e□hibit this migration pattern (e.g., *Penaeus mododon, P. notialis*) (Bhujel, 2011).

The tremendous growth rate of aquaculture has accelerated in response to decreases in commercial fish and shellfish harvests from wild fish and shellfish stocks. Each type of aquaculture system has its own set of conditions, and each system produces a different amount of total biomass. The primary criteria used to select an aquaculture system are the location, type of water body, target aquatic species, accessibility to resources and qualified personnel, and seed availability for the target species at the desired stocking time. Several socio-economic factors are also taken into account. In terms of species diversity, production in terms of quantity and value, and market share, fish constitute the vast majority of all aquatic organisms used in aquaculture. Fish culture can be done in almost any type of water, including freshwater, brackish water, seawater, and sewage water. To generate more biomass, one fish species may be chosen for aquaculture, or different combinations of fish species that get along well may be used (Meade, 2012). Traditional community ponds allow for fish stocking without adding artificial food or maintaining the water body.

1.10. PROBLEMS OF AQUACULTURE

1.10.1. Environmental Problems

Nothing is given away in the food system. Every effort to increase human food supplies has an unavoidable environmental impact. These effects on aquaculture are only now being recognized. Reporters covering the aquaculture industry should be familiar with its environmental impacts. The following are some of the most essential environmental factors:

1.10.1.1. Forage fish impacts

Many aquaculture species, particularly newly domesticated species like salmon, need wild fish rendered as fish meal and fish oil for nutrition, even though some are raised on vegetarian feed. This means that, on average, more wild fish must be consumed to produce one kilogram of farmed fish. The most well-known publication of this discovery was in the journal Nature in 2000. According to the report by Naylor et al., it could take more than 3 kilograms of forage fish to produce 1 kilogram of edible farmed fish, or "FiFo," as the feed-to-edible product ratio is known. Since the publication of their paper, the fish farming industry has become more feed-efficient. Still, subsequent research, such as "Little Fish, Big Impact" by the Lenfest Forage Fish Task Force, maintains that overfishing forage fish worldwide endangers marine food webs. Forage fish now account for about 20% of the global catch, with the vast majority being used as aquaculture feed (Bhujel, 2011).

1.10.1.2. Pollution

Fecal matter and leftover feed are byproducts of farming marine fish, crustaceans, and even bivalves. These primarily nitrogen-based wastes have the potential to reduce marine productivity and deplete oxygen levels in coastal environments. Furthermore, the use of pesticides, antibiotics, and antifoulants in aquaculture has the potential to harm the marine environment (Figure 1.5).

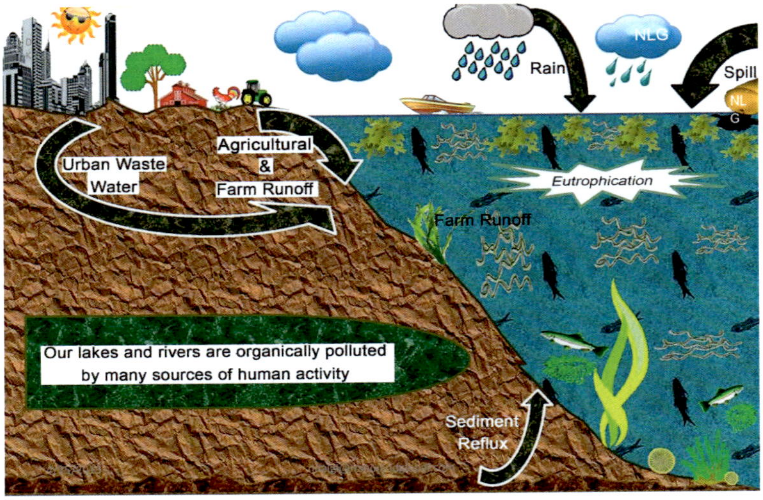

Figure 1.5. *Pollution of pesticides in marine aquaculture farms.*

1.10.1.3. Habitat destruction

Shrimp farming is likely the most controversial aquaculture practice. Shrimp are raised in tropical and subtropical ponds and impoundments, frequently located within the boundaries of coastal mangrove forests. Ponds were often abandoned and replaced in the early stages of shrimp culture because pollutants could accumulate in them over time (Meade, 2012).

Ponds were abandoned due to diseases such as viral "white spot" and bacterial "vibriosis." Hundreds of thousands of acres of mangrove forests were destroyed due to this process, which was critical for the growth of wild fish and the defense of the coast against storm surges. Mangrove cutting has recently decreased in some areas, most notably in Thailand, where it is now prohibited, but new problems have emerged. Early Mortality Syndrome (EMS), a novel shrimp disease, first appeared in China in 2009 and rapidly spread throughout Southeast Asia. In 2013, EMS-related losses exceeded $1 billion.

1.10.1.4. Escapes

Many aquaculture operations have been able to adequately address the problem of farmed fish escapes recently. Ecologists are concerned that farmed fish and shrimp are frequently genetically distinct from fish inhabiting the nearby environment, which is thought to contaminate the genetic diversity of wild populations. This is particularly concerning for the salmon aquaculture industry, as millions of fish escape from net cages suspended in the open ocean each year.

Tilapias, raised in freshwater ponds, tend to take over watersheds if they escape (Bhujel, 2011).

1.10.1.5. Disease transfer

Breeding animals in natural settings can serve as a vehicle for disease transmission. The majority of reports on this phenomenon involve disease transmission in salmon aquaculture. Infectious salmon anemia first appeared in Chile in the 1990s, and it has since been discovered in a variety of environments around the world. Poor biosecurity and the international movement of salmon larvae hastened the disease's spread from one country to the next and even from one continent to the next. The presence of sea lice, primarily of the genera *Lepeophtheirus* and *Caligus,* is another frequently reported side effect of aquaculture. Sea lice feed on their hosts' bodies by attaching themselves to the skin of their intended victims. This is especially harmful to young salmon. According to aquaculture opponents, large farm operations near wild salmon migration routes can contribute to sea lice congregating and spreading from farm to wild animals (Meade, 2012).

1.10.1.6. Managing water quality

Diseases are linked to water quality management. Because aquaculture is centered on water, the operation's success is dependent on decisions regarding the quantity and quality of the water supply that will be used. Temperature, salinity, hardness, dissolved oxygen, ammonia, nitrite, and pH are all important environmental factors for most species. Most operations strive to reduce water use and waste production while keeping these water quality factors within ranges that promote maximum development. Diet directly impacts water quality because unconsumed food and poorly digested food contribute nutrients to the water. These fertilizer inputs have the potential to fertilize both the operation's effluent and the water in which the fish swim.

These are just a few challenges the aquaculture industry will face. Aquaculture, on the other hand, will remain popular in many countries due to the rising demand for commercial fish products. Even though this industry has flaws, there have been improvements, such as what DELOS did to alleviate aquaculture concerns (Bhujel, 2011).

Although the producers were the primary cause of the problem, there have been cases in the United States where environmental degradation was so severe that the producers could claim financial losses. At the same time, ta□payers were forced to foot the bill for the clean-up. Aquaculture can also harm the environment in the following ways:

- Reduced biodiversity
- Decreased fertility of wild fish
- Incorrect reports of population growth caused by escapes
- Eutrophication of local waters as a result of increased effluent
- The extinction of sea mammals and birds as a result of aquaculture
- Depletion of pelagic fish supplies used to make fish meal
- Loss of environmentally sensitive areas due to land conversion
- Wild stock that has been genetically modified
- The spread of disease and parasites from farmed fish to wild species

Furthermore, fish may contain PCBs, dioxin, mercury, dyes, and antibiotics due to contaminated water and human input. The same fish may be toxic to one person but not to another. The consumption and production of seafood from various genera, families, classes, orders, and species necessitate the regulation of health claims and the issuing of warnings as needed. Fish that are net consumers of commodity crops, many of which are GMOs, have questionable health characteristics that may make the fish unappealing or undesirable to many domestic and international populations. While some fish may contain contaminants, there are also many benefits to consuming fish, such as high protein content, omega-3 fatty acids, and other essential nutrients that can contribute to a healthy diet. Additionally, proper cooking methods and sourcing fish from reputable sources can help mitigate potential health risks associated with contamination.

Current aquaculture methods will become obsolete if a systems approach is not adopted or aquaculture systems become overly reliant on inputs, as in land-based agriculture. For example, when rearing predatory fish, the widespread use of pelagic fish (fish that mature quickly, spawn, and die young) in the fish-in-fish-out system stresses natural systems and works against the benefits of aquaculture.

Aquaculture has the potential to provide several benefits, but these benefits can only be realized through wise resource management backed by just and enforceable regulations.

1.11. RISKS AND THREATS OF AQUACULTURE

Aquaculture risks and hazards are classified as occupational, environmental, food safety, and public health. Other elements, such as biological and chemical components, are present in all hazards (Meade, 2012).

1.11.1. Occupational Risks and Hazards

Diverse aquaculture workplaces each have their own distinct characteristics. Aquaculture risks can be classified as physical, chemical, or biological.

- **Physical:** The aquaculture industry is vulnerable to various physical risk factors. While performing their duties, aquafarm workers, including farmhands, are prone to numerous accidents. According to Clarke (2002), governments in developing countries are indifferent to occupational health and safety issues, putting fish farmers in the informal sector at risk. Farm management, employees, and governments fail to recognize the issues that occupational safety and health can resolve or mitigate. Here's a list of some physical dangers:
- **Noise:** Feedmill workers, particularly those in developing countries who use locally manufactured equipment, are subjected to excessive noise. Noise, according to Ojok (1995), has the following negative characteristics: hearing issues, hearing loss, and mental exhaustion.
- **Injuries:** Farmers are vulnerable to various injuries, including stings from fish spines. This happens when you handle fish without the proper safety gear. It can cause tetanus or witlow, as well as excruciating pain.

Cuts, sprains, fractures, and so on: these injuries can be caused by sharp implements or objects such as knives, oyster shells, falls, or other risk factors. Workers in hatcheries are also at risk of needle sticks, which can spread various viruses and diseases.

Occupational rhinitis and asthma: Feed mill workers are at risk of developing these conditions. According to Karkkainen (2002), the agricultural and food industries are the most dangerous. He says the second most common cause of asthma is animal feed and flour mill dust.

Workers in earthen pond fish farms are at risk of snake bites, crab claws, and fish bites (such as tiger fish, snapper, and so on), especially if they are not wearing the proper protective gear. Rural fish farming is an e□ample of this.

- **Mechanical injuries:** These occur in laboratories and processing plants.
- **Chemical dangers:** Aquaculture practitioners are exposed to chemical risks via the following pathways:
- **Constant use of chemicals:** Inorganic fertilizers are included and frequently used to enrich fish ponds. Formaldehyde, lime, and pesticides are among the others. Some of these are caustic, which

can cause severe skin irritation or burns, leading to occupational dermatitis. Some laboratory chemicals can be hazardous. Asthma, bronchitis, and other respiratory conditions can result from inhalation. Direct contact with these chemicals may cause burns, skin irritation, and allergies. According to observations, long-term exposure to organic solvents such as chlorinated hydrocarbons, alcohols, ester, ketone, and others puts laboratory workers at risk of brain and nervous system damage. Symptoms include premature aging, memory loss, mild depression, and anxiety. Furthermore, Karkkainen (2002) linked formaldehyde poisoning to allergic dermatitis, asthma, and rhinitis.

- **Acute and chronic waterway pollution:** Pesticides, oil spills, and other xenobiotics can pollute ponds and water sources, putting farm workers at risk.

To precipitate suspended clay particles, ponds are treated with flocculants (WHO, 1995). Calcium sulfate and aluminum sulfate are two examples.

To disinfect tools and holding areas, disinfectants such as formalin and hypochlorite are used.

Fumes, smoke, and soot: Workers who smoke fish or dry feed, as well as fumes from water pumps, feed mills, and other machines, are thought to pose serious health risks. These are linked to severe diseases such as cancer, asthma, and others.

Biological: Some of these include pathogenic infections and parasitic infestations.

One parasite that preys on people who wade in bare feet is leech in ponds. Farm workers in developing countries who use human and animal waste as inputs face risks from nematodes, cestodes, and other parasites.

- **Pathogens:** Fungus and other pathogenic infections, such as vibrio, are common in heavily manured ponds. People who were pricked by *Tilapia sp.* spines infected with *Vibrio vulnificus* lost their fingers, according to Charmish's (1996) research. Reports of fatal cases are also included in the 1992 revision of the NSSP Operations Manual.

1.11.2. Environmental Hazards and Risks

The environmental effects of aquaculture mediate a wide range of risks. Goldburg and Triplett (1997) classified them into the four categories listed below:

- ***Biological pollution*:** When non-endemic species are unintentionally released from aquaculture facilities into natural waters, it is thought to pose a severe threat to the environment. The spread of parasites and diseases that are harmful to the native population may result from the introduction of new species—numerous examples of exotic fish species wreaking havoc on native fish stocks. The lack of natural predators, parasites, or pathogens in the introduced species' new environment may contribute to their proclivity to outcompete the native stock. As a result of this trend, natural waters lose biodiversity (Meade, 2012).

The introduction of genetically modified organisms, such as transgenic fish, poses environmental risks. According toKapuscinski and Hallerman (1990), these fish could pose ecological or genetic risks if they escape from production facilities. According to the authors, one of the ecological risks would be that the transgenic fish would become an aggressive competitor or predator, negatively affecting critical ecological processes. If introduced or transgenic fish interbreed with the native stock, the gene pool may become diluted. Delvin et al., (1999) and Farrell et al., (1999) highlighted the dangers of transferring transgenic fish.

The spread of diseases and parasites from cages and pens to native stocks is a significant issue. Because disease testing and certification programs for animals are frequently not implemented, native stocks are exposed to non-endemic parasites and diseases from aquaculture facilities.

- ***Organic pollution*:** The effluents from aquaculture facilities are significant sources of organic pollution. The effluents, primarily composed of fish and feed wastes, contain a high concentration of nutrients, degrading the water's quality and encouraging the growth of undesirable algae.
- ***Chemical pollution*:** Chemical use in ponds and laboratories endangers the environment. These substances have the potential to be disruptive, and if they end up in natural aquatic ecosystems, they have the potential to destroy the ecosystem permanently. Environmental risk factors include fertilizers, pesticides, antifoulants (for cages), and chemotherapy drugs (Meade, J. W. 2012).
- ***Habitat modification*:** Aquaculture sites can have a negative environmental impact. Aquaculture development can occasionally change the topography of aquatic systems, destroying habitats and eradicating biodiversity. The newly created habitat may be unable to maintain ecological balance.
- ***Threats to wild species:*** Aquaculture fish that escape interact with genetically diverse wild populations, competing for resources and spreading diseases.

Carnivorous farmed fish are fed fishmeal and oils from forage fish (sardines, anchovies, mackerel, herring, and crustaceans, mainly krill). Omega-3 fatty acids, vitamins, and minerals abound in these species. Over the last ten years, the number of fishmeal factories in West Africa has increased significantly. As a result, small forage fish, a staple of the local diet traditionally caught by artisanal fishermen and smoked and dried by women, are being overfished. Food insecurity, job losses in the local artisanal sector, environmental damage, and public health risks result from fish stock depletion in areas where fishmeal plants are installed due to contamination.

- ***Ecosystem destruction:*** When kept in pens, the fish are more likely to sustain injuries, become ill, and contract parasites. To combat these effects, fish farmers use pesticides and antibiotics that contaminate the water. Wastewater contaminated with food residue, antibiotics, and excrement causes dead zones in the surrounding environment of farm sites.

Coastal ecosystems are frequently destroyed to make way for intensive aquaculture. This is true of the man-made ponds where tropical shrimp are raised. Mangrove removal results in the extinction of every species that once found refuge among the trees, including commercially valuable fish, oysters, birds, and others. This includes destroying a storm and tsunami defense system (Meade, 2012).

1.11.3. Public Health and Food Safety Risks

Aquaculture products, like other foods, may pose health risks to consumers. The manufacturing process also endangers the general public's health. Aquaculture products pose significant biological health risks, particularly for organisms originating in wastewater or receiving animal and human wastes.

According to Edward (2001), pathogenic enterobacteria (from the human digestive tract) have been discovered in fish guts. As a result, the safety of eating fish in such environments is called into question. Food-borne trematodes such as *Clonorchis sinensis* and *Opisthorchis viverrini* have been linked to illness, particularly in Asians who consume raw or undercooked fish. Trematode parasites have been linked to enteric diseases in Egypt and the Republic of Korea (WHO, 1999). Accidental infections have also been reported with nematodes transmitted by fish.

Human pathogenic bacteria found in fish and crustaceans pose a risk. These bacteria are caused by contamination from human or animal waste and local flora (Buras, 1993). Post-harvest handling and processing are additional sources of contamination. Toxic compounds are produced by aquatic microorganisms, such as algae and detritus, and pose serious health risks to humans. *Alexandrium tamarensis*, the dinoflagellate that causes to□ic red tide, is a good e□ample. People who consume contaminated shellfish (lobsters and crabs) risk developing paralytic shellfish poisoning (Bhujel, 2011).

Furthermore, shellfish have been linked to the transmission of human pathogens such as the Vibrio bacterium, which causes gastroenteritis in humans. Filter feeders such as clams, mussels, and oysters may also endanger consumers' health because they accumulate contaminants in their internal organs. The consumption of raw mollusks harvested in waters contaminated with untreated or raw human sewage is the leading cause of seafood-related illnesses, according to Ahmed (1991) and Hackney (1994).

Some chemical products used in aquaculture are considered hazardous regarding food safety. Chemotherapeutic drugs, lime, flocculants, algicides, chemical fertilizers, and flocculants are among them. Some of these chemicals may become biomagnified in animal tissue, putting consumers at risk of chemical intoxication. Other chemicals, such as industrial hydrocarbons, may leak and contaminate aquaculture water sources. This is more visible in urban aquaculture, where wastewater is recycled. According to the WHO (1999), fish raised in contaminated water should be considered potentially dangerous.

Additional risks to public health are the growth of mosquito larvae and cercaria, which increase the incidence and prevalence of malaria and filariasis, respectively. This is more common in developing countries where there are a lot of tiny fish ponds, which encourages higher densities of these organisms. Antibiotic and veterinary drug residues in food are also a concern. Furthermore, antibiotic resistance has been observed in areas where aquatic farm animals are treated. According to Angulo (2000), the multidrug-resistant *Vibro cholerae* was caused by the use of antibiotics in the Ecuadorian shrimp farming industry. It would be challenging to treat humans infected with such antibiotic-resistant organisms. It is critical to consider the contentious issue of GM fish safety. Transgenic fish have been labeled as dangerous in terms of food safety due to their potential toxicity and allergenicity. However, these claims must be supported. The introduction of pathogenic organisms during the production of products under unsanitary conditions raises public health concerns (Meade, 2012).

1.11.4. Methods for Managing and Controlling Risks and Hazards in Aquaculture

The guiding principles for risk management in aquaculture are identification, mitigation, and monitoring of the effectiveness of controls. This section has listed the risk factors and hazards in the preceding section. This section would recommend control measures to reduce or eliminate aquaculture risks. As previously stated, workers' ignorance and the indifference of employers and government organizations to their plight have resulted in avoidable fatalities. Governments, businesses, and consumers all play a role in ensuring the production of safe aquaculture-based foods, as each plays a vital role in preserving human health. The development of rules, allocating resources to enforce them, education and training in responsible aquaculture practices, and related research necessitate action at all levels. The following are recommended:

Employees should receive thorough training on the hazards and risks specific to their line of work before beginning work. To instill safety consciousness in older employees, they must be reoriented.

Personal protective equipment should be available to all staff categories. The use of such equipment should be strictly enforced to reduce the risk of accidents or other workplace hazards.

Regular medical examinations of laboratory workers and other staff members who handle chemicals should be required to detect any adverse effects of chemical poisoning as soon as possible.

All aquaculture facilities should have first-aid kits on hand, as well as clear instructions on how to use them.

Accurate records of aquaculture-related risks should be kept to raise awareness of these risks and be updated. This will serve as a preventative or precautionary measure.

Establishing a specialized occupational medical clinical service with access to specialized diagnostic and management resources is critical. This service would provide immediate intervention, occupational disease diagnosis, and management.

Governments must enact proactive laws and policies, anticipating problems and implementing solutions. These measures must be implemented (Meade, 2012).

Relevant parties should provide guidelines for achieving a minimum level of environmental protection near aqua farms. Nations should use the Hazard Analysis Critical Control Point (HACCP) framework to improve food safety.

A thorough understanding of the relationship between reducing food-related hazards and lowering consumer risk is critical for developing adequate food safety controls.

A coordinated strategy involving health promotion, vector control, and targeted population chemotherapy should be used to combat parasite issues.

It is best to discourage the indiscriminate and uncontrolled use of chemicals and medications to protect the environment and the general public's health.

Every human endeavor contains the potential for risks and hazards to affect individuals. This is also true for aquaculture. However, the available information focuses primarily on dangers and risks in developed countries. Unfortunately, policies (where they exist) that could ensure safe aquaculture practices in developing countries, where 87% of the world's aquaculture production is produced, are frequently ignored by workers, employers, governments, and consumers. Some countries still need policies to ensure regulatory standards are followed. As a result, aquaculture has become unregulated, which has had negative consequences (Bhujel, 2011).

1.12. POTENTIAL OF AQUACULTURE

According to a recent UN study, aquaculture can increase the amount of seafood available for consumption, thereby improving food security and nutrition. Aquaculture can increase food production, stimulate rural and coastal economies, and clean waterways.

1.12.1. Shellfish Beds Aid in Waterway Restoration.

Shellfish such as oysters, clams, and mussels are popular at beach barbecues and play an essential role in waterways. However, these shellfish have virtually vanished from coastal ecosystems in some areas. By restoring and farming shellfish beds, these bivalves can clean the water by filtering feed, acting as a habitat for smaller organisms, and even assisting in preventing shoreline erosion. Protecting biodiversity, preserving ecosystems, and halting land degradation are all critical components of sustainable aquaculture (Meade, 2012).

1.12.2. Increasing Seafood Production can Reduce Food Insecurity.

Despite covering 70% of the Earth's surface, the oceans produce less than 2% of global food production. Fortunately, more sustainable aquaculture farms are emerging to meet our growing population's needs. Aquaculture is becoming known as one of the most environmentally friendly ways to produce food and protein when practiced ethically, as it is in the United States. Leading health experts emphasize the importance of including seafood in one's diet, especially for those who are pregnant, nursing, or have young children. Farmed seafood can improve nutrition and food security in many communities (Bhujel, 2011).

1.12.3. Aquaculture as a Tool for Economic Development

Aquaculture helps to build resilient working waterfronts, and coastal communities provide year-round employment and have the potential to generate long-term economic growth. Aquaculture production in the United States is worth $1.5 billion. Aquaculture impacts employment in other industries, such as feed and equipment manufacturing and seafood processing.

1.12.4. Aquaculture that is Grown Sustainably Employs Advanced Technology.

Aquaculture farming techniques that are environmentally friendly prioritize food production and environmental protection. OceanReports, a free mapping and planning tool that e□amines "ocean neighborhoods," is an e□ample of a modern device. It provides critical information for making sound decisions that increase sustainable aquaculture opportunities while minimizing potential environmental consequences. It also provides detailed information about local industries, species, habitats, and potential aquaculture risks (such as shipwrecks and undersea cables) (Figure 1.6).

Figure 1.6. *Using sustainable aquaculture to increase our fresh seafood supply can boost the ecosystem and the economy.*

Credit: NOAA Sea Grant.

1.12.5. Eating Fish, Both Wild and Farmed, can be Beneficial to Your Health.

Seafood is not only delicious, but it is also good for your health. It contains nutrients that have been linked to a variety of benefits, including a lower risk of coronary heart disease. Shellfish and fish, particularly salmon, tuna, and trout, contain omega-3 fatty acids, which promote heart and brain health. Because our bodies cannot produce omega-3s, they must be obtained through food; therefore, reliable sources of these fats are essential. As aquaculture continues to increase seafood production, more people will be able to include seafood in their diets, improving nutrition and well-being (Meade, 2012).

1.12.6. Aquaculture Provides a Chance to Preserve Cultures.

Traditional Hawaiian fishponds (loko i'a), which have been used for nearly a thousand years, are a treasured Pacific Island cultural and environmental tradition. Many of these fishponds, used by native Hawaiians to raise and fatten fish until they were ready to be harvested, have unfortunately deteriorated. However, an increasing number of pond managers are working to restore these ponds, and the revitalized areas are becoming essential parts of the community. As more fishponds are fixed, the availability of local seafood, employment opportunities, and other revenue sources may increase. The fishponds promote economic growth, lucrative employment, and sustainable resource management (Bhujel, 2011).

1.12.7. Sustainability Promotes Global Cooperation

Sharing expenses, goods, tools, data, and technologies benefit marine aquaculture partnerships. International aquaculture collaboration creates a large scientific community that exchanges ideas, cutting-edge aquaculture techniques and management strategies, and emerging trends. Countries can achieve their marine stewardship goals while improving seafood access by exchanging and implementing cutting-edge sustainable aquaculture techniques. Collaboration allows us to work together on ocean use, advance sustainable development, and develop innovative technology.

1.12.8. Aquaculture is a Future-Oriented Food Production System.

Today, there are more than 7 billion people on the planet, and by 2050, that number is expected to rise to close to 10 billion. Traditional land-based agriculture takes up more than half of all developed land and more than 70% of the world's freshwater resources (Meade, 2012).

Although the oceans cover more than 70% of the planet's surface, capture fisheries and a small marine aquaculture industry produce only 2% of the world's food. Ocean fish farming has the potential to significantly reduce the strain on the planet's limited natural resources. Partnerships allow countries to pool resources and move forward more quickly.

Aquaculture has many environmental advantages over other livestock farming methods. Marine aquaculture operations typically have a lower carbon footprint and require less freshwater and land. They are more efficient than beef, pork, and poultry at converting feed into protein for human consumption. Sustainable aquaculture must be expanded to increase food production in a changing environment.

REFERENCES

1. Agarwal, A. (2019). *Getting the Shrimp's Share: Mangrove Deforestation and Shrimp Consumption Assessment and Alternatives*. Retrieved from: https://www.iddri.org/sites/default/files/PDF/Publications/Catalogue%20 Iddri/Rapport/Getting%20the%20shrimp%20s%20share.pdf (accessed on 3 August 2024).
2. Mark, R., Junzo, E., Keisuke, M., Juzhong, Z., & 000, Y. (2019). Common carp aquaculture in Neolithic China dates back 8,000 years. *Nature Ecology & Evolution, 3*, 1415–1418. https://www.nature.com/articles/s41559-019-0974-3 (accessed on 3 August 2024).
3. Martinell, D.P., Vergara-Solana, F.J., Padilla, M.E.A., & Garza, F.A. (Eds.). (2024). An Introduction to Sustainable Aquaculture (1st ed.). Routledge. https://doi.org/10.4324/9781003174271
4. Orville, D., & David, K. (1979). *Mariculture in New York State*. Retrieved from: https://repository.library.noaa.gov/view/noaa/38505 (accessed on 3 August 2024).

Chapter 2

An Overview of Aquaculture Systems

Contents

INTRODUCTION

No matter how advanced aquaculture technology grows, it will not be sustainable unless it is backed by laws that take into consideration the social and economic context in which the aquaculture system operates. Environmental regulations are required for the advancement of such technologies.

Beginning sometime around the year 2015, it is anticipated that water shortage would emerge as a major problem in the twenty-first century. The high demand for this essential component of aquaculture will make it difficult to carry out ordinary operations. Because of the widespread drinking water shortages that are expected to occur by the year 2025, there will likely be increased competition for this resource. Because of this underlying issue, the spread of aquaculture will face substantial challenges throughout the course of the ne□t millennium. In order to find a solution to this problem, appropriate agricultural methods and technological advances will be required.

In the context of this chapter, the requirement for ownership, care for aquatic stock, confinement or site selection, varying degrees of isolation from the outside environment, allowing for varying degrees of internal system control, and ownership are all considered to be essential components of aquaculture. A contract to that effect may also be considered an essential component of aquaculture.

When evaluating aquaculture systems, natural resource management and the environmental context in which the systems are located should both be taken into consideration.

It is essential to have a broad perspective and think about how aquaculture fits in with other activities and procedures from an economic and social point of view. Since aquaculture may have an effect on other users of water and natural resources, and other users can have an effect on aquaculture, it is essential to understand how the two are related. Aquaculture has an effect on other users of water and natural resources.

Aquaculture methods could be general, semi-intensive, intense, or even hyper-intensive. Because there are no clear divisions and the intensification levels reflect a continuum, it is essential to provide a comprehensive definition of each system when using this word.

There are several approaches to farming, some of which include: water-based systems recycling techniques land-based methods depending on agriculture.

Echinoderms, coelenterates, seahorses, and other minute invertebrates such as mollusks and crabs are all examples of invertebrates that incorporate numerous aspects of agriculture into a single system. Examples of marine life include many types of fish as well as macrophytes and seaweed. The process of caring for broodstock, the creation of seeds in hatcheries, nursing systems, grow-out systems, and quarantine are all included in aquaculture. Because of the wide variety of culture intensities, culture systems, species, farming methods, and culture phases, aquaculture technologies and systems are incredibly diverse. This is due to the fact that aquaculture production can take place at varying stages.

Aquaculture includes a wide range of management interventions, infrastructure, and related technologies. These include the provision of seed, stocking, handling, and feeding, as well as the regulation, monitoring, sorting, treating, harvesting, and processing of produce.

Aquaculture is influenced by a wide variety of causes, only two of which are human requirements and economic considerations. Both technology and human-based solutions are necessary in order to develop a sustainable aquaculture industry. They are able to learn about a variety of cultural systems and select one from among those systems until they find one that fulfills their needs and accounts for the advantages and disadvantages of the environment in which they live. Even if these methods are currently in use, it is imperative that more people become aware of them. In order for them to accomplish these objectives, they require trustworthy information on the benefits and downsides of various approaches, as well as assistance with the decision-making processes that are involved in the process of building up production systems.

In many parts of the world, there will soon be a water shortage, which will cause anxiety. However, when the weather is hot and dry, a pond that is one hectare in size might lose as much as 30,000 tons of water every year as a result of seepage and evaporation, while only producing one to two tons of fish. Using "intense" quantities of water is necessary for ponds. It is possible for aquaculture ponds to be integrated into water management and conservation systems, which would make rain-fed aquaculture a practicable alternative to water storage in dry regions.

This suggests that a significant number of farmers will have a need for education, direction, and the speedy distribution of innovative aquaculture methods that make the most use of restricted water supplies.

Flow-through systems may be made more efficient by the recycling of thermal energy, which can also assist in keeping the cost of water consistent in certain locations. The incorporation of pond systems, the reuse of water, and the recirculation of water are all practices that result in good impacts. For every kilogram of fish, for instance, a de-nitrification machine may generate 40 liters of water, whereas a recirculating system may produce 150 liters of water. The problem is that the vast majority of aquaculture producers could discover that it is too difficult to implement these methods. It is possible that many of the "outputs" of agricultural subsystems, which are more commonly referred to as "waste" or "byproducts," are really essential inputs for other subsystems, rather than merely being additive components of the agricultural economy as a whole.

There are several instances of relationships that are structured in such a complex web. In Israel, for instance, there are systems that store irrigation water alongside aquaculture ponds and move water on a regular basis to suit the different requirements of the two purposes. These twin pond systems are called "twin pond systems." Small-scale integrative activities may be made possible in ponds and reservoirs thanks to cages, which leads to an increase in the economic efficiency of water resource usage. Rainwater aquaculture ponds are a more effective way of putting this resource to use than other methods. Aquaculture production is an added advantage. Aquaculture is favored by certain other types of smaller-scale systems, with water storage acting as a backup.

Despite the fact that livestock and aquaculture are commonly intertwined, there may be ethical questions raised over the usage of livestock or industrial waste products in the future. The spread of infectious diseases and the accumulation of hazardous substances are two separate but related concerns. Due to variables such as Bangladesh's plans for flood management or the utilization of inappropriate ground or surface water for agricultural or human use, alternative water sources for aquaculture are readily available and may be used without much difficulty. Aquaculture is a realistic choice since many Asian nations have huge sections of salty soil and water, which allows for the practice.

Recirculation has a variety of purposes, including breeding stock conservation, growing out, and quarantine. Recirculation systems are expected to become more prevalent in enhanced commercial aquaculture in the future. There are several possibilities that may be tailored to local conditions.

The American catfish PAS system is one e□ample. A thick network of pond-internal tubes for water filtration is combined with a massive growth cage. The extremely slow speed and low energy demand of the circulation efficiently regulate environmental processes in the pond. Water recirculation does not always need a lot of effort. Closed pond systems are used by Thai shrimp farmers to reduce the requirement for water exchange, optimize the use of brackish water, and reduce the risk of disease introduction into the shrimp farming system.

Active suspension ponds, which reduce the requirement for water exchange, have been shown to be beneficial for tilapia and shrimp production in Israel, the United States, and Belize. There are several advantages to having extensive recirculation systems. These qualities include low water requirements, restricted space requirements, reduced water discharges, closely regulated feeding to optimize feed conversion efficiency, generally site independence, e□clusion of predators and climatic events, and little chemical usage. However, these systems are more complicated, have greater capital costs, and failures can result in significant crop loss, in addition to the use of regulated settings to enhance productivity and feed conversion efficiency. Such systems need professional management and place a greater emphasis on feed design, management control, and health management.

Given that the vast majority of current applications are for high-value intensive aquaculture, a well-designed recirculation system must be both simple to operate and cost-effective. Because of environmental concerns and the market for highly valued aquacultured species, Europe is presently best positioned for hyper-intensive recirculation. Alternative recirculation applications are expected to emerge as future resource and economic conditions shift.

Concerns concerning technology recirculation techniques include the following:

Recirculation devices confront several technological challenges, including a lack of knowledge of component interactions. The interactions in biofilters between pathogens and beneficial microorganisms are poorly known. More study is needed on biofilms, biomats, and other biomaterials.

Scale-up concerns are common and need rigorous testing. When using fresh items, processes may need to be changed. The buildup of byproducts in the system is poorly understood. Predictive modeling is required for evaluating multifactor interactions in the design and testing of recirculation systems.

The efficacy of water treatment must be matched with the efficiency of conversion when constructing feeds for recycling systems. Feeds are now being developed to help separate feces from water and prevent nutrition loss.

Because they allow for more efficient management of wild escape, recirculation systems are preferred for the growth of foreign species and GMOs.

Intensification can disrupt fish social processes, causing them to become stressed. This depends on the species since some can adapt to massive stock levels. Prior adaption to a recirculating environment may be required for fish. Techniques for recirculation vary depending on the species. It is feasible to choose species that are difficult to produce yet perform well in recirculating systems. Strains developed and adapted for recirculating systems appear to work as planned.

We must design systems that are adapted to the needs of farmed animals out of concern for their well-being and the need for higher productivity.

Energy savings may be the limiting factor that makes recirculation appealing in some cases, such as heated hatchery and/or grow-out systems.

The ocean, particularly offshore places, will be an important habitat for the future development of aquaculture. Coastal waterways, bays, and inlets, among other places, are now in use; nevertheless, open water development can be too expensive at times.

As we approach the twenty-first century, it is clear that the global aquaculture production growth rate is slowing. If this is due to manufacturing constraints, it shows that we are not making the most of the technology we now have; otherwise, future advancements will be more difficult to achieve. Basic technological advances in aquaculture are required to improve culture conditions. It would also be beneficial to investigate the possible performance of other species.

The increase in aquaculture output has slowed dramatically due to saturation, disease difficulties, environmental constraints, and the effect of large current producers. We should also think about the great long-term opportunities in South America and Africa, where adequate technology may already exist but has not yet been broadly disseminated in a way that is acceptable for local conditions.

The global production of caged fish is increasing. Technological improvements have been great in China, South America, and certain parts of Europe. The fast development of fish cage farming in Southeast Asia is slowed by a lack of seed, as well as the high cost and limited volume of food. Fish fingerlings and seed are already being transported across Asian borders, mostly to live fish markets in China and Hong Kong. Although it is acknowledged that this trade resulted in harmful fishing tactics for fish fingerlings, nothing is known about the environmental consequences.

Future cage culture expansion is projected, but it must take into consideration the species, markets, and obstacles unique to each location.

The ecological impact of inshore and near shore cage farms can be either favorable or negative. Combining fish cages with seaweed and shellfish culture, for e□ample, may reduce nutrient and organic loading; however, combining fish cages with artificial reefs can boost stocks and, in the long run, may result in culture-based fisheries.

Concerns about environmental management will be critical for the future growth of cage culture. The following are the concerns that must be addressed: A better understanding of mortality and the actual number of fish in cages; improved feeding regimens with less feed waste; thorough investigation of material and energy flows through cage systems; modeling of environmental impacts; better understanding of recovery processes in order to estimate fallowing time; site rotation: improved equipment for simpler mooring; better management of coastal zones, access rights, and a reduction in the use of toxic chemicals.

In theory, open-ocean aquaculture is safe, improves effluent e□change and dispersion, and necessitates enhanced monitoring and remote feeding management. The vast majority of extant instances are still prototypes, and manufacturing costs remain unreasonably high. Furthermore, less is known about the performance of fish offshore vs. inshore.

Offshore cage system infrastructure needs may vary. Systems may be left offshore for lengthy periods of time without being hauled in for maintenance; net repair and maintenance will be handled separately. Other support systems, such as sorting, harvesting, processing, inspecting, and treating space, may be required offshore.

The medium-term applications of this technology will focus on high-value species.

The following infrastructure and support technology issues are associated with environmental assessment and planning:

1. Aquaculture computerization;
2. Financial control;
3. Decision support tools for planning;
4. More effective farm management software;
5. More reliable online monitoring devices, in addition to calibration and self-diagnosis.

More effective zoning, regulation, and environmental management. GIS integration into system management. Improved impact assessment methodologies, such as carrying and holding capacity models. Widespread use of vaccinations and parasite treatments in the vicinity of manufacturing enterprises.

Aquaculture is not subject to the same material constraints as other food industries. Polymer components with low molecular weight, for example, might be a source of contamination. Flame retardants, plasticizers, stabilizers, lubricants, pigments, UV absorbers, and antistatic agents are some examples. Criteria for materials in recycling systems may be necessary in this scenario, as items entering the solution have the potential to poison the system and the end outcome.

Despite the fact that some countries have exceptionally rigorous aquaculture feed quality and standards, many countries will need to become even more severe in order to fulfill e□port and market demands.

We may appear to be seeking to develop an excessive number of species compared to the cattle sector. On the one hand, because there are so many different species, we can employ a range of settings and niches to boost aquaculture output. Alternatively, while we currently govern the cattle industry's limited possibilities, the variety makes consistency and generalization difficult.

When picking species, keep in mind consumer demand, the availability of seeds and production techniques, and any potential environmental constraints. Diversification is important in Mediterranean aquaculture because numerous fast-growing species may share infrastructure and have short life spans.

An increase in seed supply; air-breathing fish; situations characterized by a lack of technology or poor water quality. More microalgal species are required for the generation of fine chemicals and hatchling feeding. In tropical and subtropical climates, similar assertions are made about alternative shrimp and crustacean species, indigenous fish species as opposed to Chinese and Indian cyprinids, freshwater mussels, shrimp, and snails.

The idea is to abbreviate the life cycles of the animals that are now farmed in aquaculture.

The discovery of innovative products that may be used for purposes other than human consumption, as well as the extension of the usage of currently grown species, may greatly increase aquaculture's economic efficiency. As an e□ample of the use of fish as a bioreactor or instrument in pharmaceutical research, salmon may be utilized to treat osteoporosis.

Education and training must be prioritized in the new millennium. Course accreditation, competency standards depending on the sector, and frequent refresher training for training providers should all help to assure the quality of training.

It is critical to create innovative techniques for efficiently presenting information to decision-makers, government personnel, investors, engineers, journalists, and members of the general public who may have an impact on aquaculture expansion.

2.1. POND AQUACULTURE SYSTEM

This technique dominates the fish farming industry. Fish and shellfish are raised in ponds that have been dug out specifically for that purpose. The water for the ponds may come from a bore well, a bore canal, or even just from rain. Now that the best spot has been selected, it's time to dig the pond. Fish pond conditions are affected by factors such as local climate, terrain, water availability, and soil quality.

The following characteristics separate the two kinds of ponds: natural ponds found inside watersheds

1. Lake Levee secondly, a network of ponds that serve as a watershed.
2. Watershed runoff is the only source of water for watershed ponds; groundwater and surface water can be used as backup supplies.
3. Soil type, features, geography, and water supplies are the most crucial factors.
4. The water retention capacity of ponds is affected by the soil type used to build them; mud and mud-silt soils are preferred since they do not leak. It is important to use high-quality soil that contains no more than 20 percent clay while constructing a pond.
5. The size and shape of ponds are influenced by the surrounding landscape. No pond in the watershed should be deeper than 10 feet. The recommended maximum size for a pond is 20 acres. A large pond site can be split into multiple smaller ponds.

Figure 2.1. *Shrimp ponds.*

The location and type of pond to be constructed depend on the available water and its amount and quality. The benefits of this pond system include cheap pond building, no water expenses, and less rivalry for water from other agricultural applications, however, the challenges of growing and harvesting in watershed ponds outweigh these advantages (Figure 2.1).

1. Dams and Levees for Water Features

When the water supply from the watershed is insufficient, levee ponds are constructed on flat land. A lot of times the ponds become flooded because of the groundwater. Verify the area is suitable for a pond by inspecting the soil, water, and groundwater. The lifespan and capacity of ponds are affected by the soil type they are built upon. The soil around the pool has to be 20% clay in order to prevent leaks. Location dictates pond size; rectangular ponds are preferred for ease of maintenance, feeding, and harvesting. Because of the challenges they provide in maintaining water quality, ponds that are curved or uneven are not advised. The condition of the water must be checked before any work begins on the pond. The watershed pond is not as good as Levee Pond for fish farming.

2.2. PEN AQUACULTURE SYSTEM

Aquaculture is the production of fish and other marine creatures for human consumption. It is possible to breed fish in both freshwater and saltwater environments by following this approach. The cultivation of fish in man-made environments such as ponds, tanks, pens, or cages is known as aquaculture.

If they are kept in a safe environment and the water is not tainted in any way, pens and cages have the potential to be an appropriate environment for the development of fish. The use of pens and cages is the most common method of fish farming in India. Fish farming in pens and cages is a relatively new concept in Nigeria, but it holds a great deal of potential for the country.

1. Pens

Permanent enclosure in an open body of water constructed with bamboo or wood poles and netting to allow contact with the surrounding environment but preventing unwanted fish and animals from entering and damaging the provided fish (Figure 2.2).

When constructing a fish cage, keep the following elements in mind:

- *Location*

The enormous ponds or other bodies of water that mimic ponds where pens reside are often found far from shore. This method is environmentally friendly since it has no negative impact on the environment.

- *Construction*

The structure of the pen is made of bamboo or wooden poles and is subsequently submerged in the lake or buried on the coast. Nets are stretched from pole to pole and weighted with lead or heavy stones to make the cage for the dishes.

Figure 2.2. *Schematic of pen aquaculture system.*

2. The advantages of keeping fish in tanks

The following are some of the numerous advantages of using pens in aquaculture:

- Commercially bred fish mature at a breakneck pace in nutrient-rich lakes.
- This strategy works just as well when used to unproductive water by adding additional food to it. Pen culture is also a popular form of fish rearing since it has no detrimental impact on the natural environment.
- Pens should be replaced every year because they only last three to five years. Because fish cages are usually located in shallow water where fish reproduce, they may diminish natural production in certain lakes. To reduce rot and water gathering, nylon or plastic netting is advised.

3. Cages

Cages are a type of floating confinement that is commonly employed in fish farming. Cages are commonly used in commercial aquaculture. Cage-reared fish are demonstrated to be capable of reproducing a wide range of fish species. Cages come in a variety of sizes and designs. Bamboo and other locally accessible, low-cost materials, for example, can be utilized to construct cages. Cages can instead be made from PVC tubes. Cages are adaptable enough to be employed in both marine and freshwater settings, offering enough protection against natural calamities and human intervention.

4. A diverse variety of cages
   - A pen that can float. It is helpful in bodies of water at least 5 meters deep.
   - Crate that remains stationary: This cage's optimal water depth is less than 5 meters.
   - A cage is shaped like a rectangle in its most basic form.
   - A cage in the form of a hoop

If the cages were fashioned from locally obtained bamboo, they could be made more inexpensively. PVC pipes may also be used to make cages. Cage bodies with the required sizes and configurations may be readily constructed using bamboo and nylon ropes, as proved in prior tests. A tightly knotted inner hapa or nylon fabric cage with a closeable cover is housed within the bamboo framework. The mesh size of the bamboo-woven main cage and the size of the nylon inner cage are critical to cage fish farming success. The bamboo cage and the nylon liner within must be cleaned on a regular basis in order for the fish to live. These cages are made of matured bamboo and last three to four years, whereas those made of seasoned and mature bamboo endure eight to ten years.

The benefits of cage culture there is a large fish output, and private ownership inside publicly accessible rivers is possible, with easy management of fish-specific predators and competition. Storm damage, pollution, theft, and vandalism are all problems associated with cage culture, and mounting security costs may jeopardize the practice's long-term viability.

2.3. CAGE AQUACULTURE SYSTEM

Cage culture is currently receiving more attention from researchers and groups. Caged fish farming has grown in popularity as a result of factors such as rising fish consumption, dwindling wild fish populations, and bad agricultural economics. Many small farmers and farmers with limited resources are looking into alternative crops to replace traditional agricultural products. Despite its small size, aquaculture has the potential to be a rapidly growing industry. Furthermore, cage cultivation allows farmers to use water sources that are typically underutilized for other uses.

The location of a cage farm has a significant impact on its productivity. Location is critical because it may have a significant impact on economic viability by influencing capital e□penditure, operating e□penditures, output rate, and mortality characteristics.

The location of an aquaculture facility has a direct impact on its success and sustainability. Shrimp, crabs, and lobsters, as well as milkfish, mullet, cobia, pompano, sea bass, and pearl spot, have all been successfully cultivated in circular cages ranging in diameter from 2 to 15 meters, as tested and demonstrated in India at the Central Marine Fisheries Research Institute. Stocking needs and ranges vary per fish species.

The proper feeding of animals, frequent inspection and cleaning of their cages, and the maintenance of hygienic conditions are all critical to the success of cage farming. A cage in the best possible site may generate 20 to 40 kg of fish per square meter, although this varies by species.

Fish are raised in their natural environment, which is a body of water, in a net cage that does not obstruct water flow. A reservoir, river, lake, or ocean can all serve as places for an aquaculture production system that uses a circular or square floating net coupled to a frame that floats on the water's surface and a mooring system to store and produce a huge number of fish. A catwalk and fence encircle a number of cages dangling in the air. The four major forms of fish farming are fi□ed cages, floating cages, submerged cages, and submersible cages. Because of its low carbon footprint and great yields, cage culture is a sustainable agricultural approach. The necessity for a regular supply of sterile, o□ygenated water is one of the most difficult issues of land-based fish farming, which is overcome by raising fish in an e□isting body of water. Fish in cage farms deliberately located to take advantage of natural currents are subjected to higher levels of oxygen and other natural conditions (Figure 2.3).

Figure 2.3. *Cage system in a fish pond.*

Because of their high productivity, cage culture methods have the potential to significantly increase the total fish supply of India. India has a long coastline and numerous huge areas of brackish water accessible from coastal states, both of which might be perfect for cage culture. Because of the minimal or non-existent land requirements and cheap overhead, this farming approach is suitable for recreational fishermen. Because of the modest amount of labor required and the manageability of a small home, this may be a feminine or domestic task. Modifications to the cage and its accessories can be made to meet the individual needs of each farmer.

1. The benefits and drawbacks of cage confinement

Before deciding to adopt cage-raising fish, the benefits and drawbacks must be thoroughly studied. An ambitious fish farmer might reduce risks by beginning small in a pre-e□isting pond that is not utilized for recreational fishing.

- Advantages:
  - o Cage farming allows for the extraction of water from otherwise inaccessible locations such as lakes, reservoirs, ponds, strip mines, streams, and rivers.
  - o Using an existing body of water requires less upfront expenditure.
  - o Harvesting is less of a nuisance, and fish can now be e□amined and sampled with reasonable simplicity.
  - o It makes the pond suitable for animal breeding and recreational fishing.
  - o There is less work to be done.
  - o Making it simpler for unemployed young people and women to obtain work; allowing fishermen to earn more money during the off-season.
- Disadvantages:

Furthermore, there are some obvious disadvantages to cage culture. It is critical to provide fresh and healthy feed.

- o Mechanical aeration has the ability to ease the persistent condition of Low Dissolved Oxygen Syndrome.
- o The netting of the cage has been damaged, and the risk of catching and spreading a disease is quite high.
- o Vandalism and poaching are two potential hazards.
- o Improper navigation.
- o Excessive feed and trash will cause eutrophication and water contamination.

- o Changes in water quality indices.
- o Too many aquatic species are caged, which causes issues such as escape and disputes with neighbors.

2. The practice of cage-keeping fish
   - Site assessment

A variety of parameters influence the site of cage farming.

Temperature, salinity, oxygen, wave action, pollution, algae blooms, and water e□change are all physical and chemical factors that influence whether or not a species can thrive in a particular habitat. Before deciding on a location, consider factors such as climate, cover, water depth, and substrate. Legal considerations, accessibility, proximity to hatcheries or fishing harbors, security, economic, social, and market difficulties are all important.

- Cage dimensions

Given the limits of the materials and procedures used, the cost per unit volume decreases as the cage size increases. CMFRI has developed open sea cages with diameters of 6 and 15 meters for grow-out fish culture, as well as seed-raising HDPE cages with a diameter of 2 meters for propagation. Because of its mobility and low maintenance requirements, grow out cages benefit immensely from being 6 meters long. Fingerlings can be securely housed in cages no more than two meters long.

3. Mesh cages and other mesh

Many different types of cages can be used in farms that confine animals. Densely packed Polyethylene, polyvinyl chloride, and galvanized iron are common pipe materials. Although pricey, HDPE frames are sturdy. Epoxy-coated galvanized iron frames are affordable and long-lasting, making them an e□cellent choice for smaller groups and fishers. GI frames have a lower lifetime as compared to HDPE frames.

In India, many different types of cage culture nets were tested. CMFRI has used tangled and braided HDPE nets for horticulture. It's conceivable that more than one season may be released. Although nylon net is a low-cost material, its light weight demands the addition of ballast to keep its shape. The usage of pricey, cutting-edge netting materials like as sapphire or dyneema raises the question of whether such advancements should be used. A net's ideal depth is between two and five meters. A predator net is required for open sea cage farming to protect against predatory species.

4. Animals that can be kept in cages and the criteria used to select which ones

When selecting a species for cage production, several biological characteristics like as omnivore or carnivore status, quick development, effective food conversion, seed availability, disease resistance, and market demand should be considered.

Despite the notion that species needs and operational problems should impact stocking levels, the effect of stocking density on growth and productivity has been experimentally shown. The eating habits of the planted species, in combination to the carrying capacity of the cages, will decide the ideal stocking density. The appropriate stocking density will be determined by the kind and size of the fish.

5. Feeding and dosing procedures

Fresh or frozen waste fish, wet pellets, and floating dry pellets are commonly served to caged fish. Cage feeding is less complicated than pond feeding. Because the ration is easily divided in half, it may be distributed on a regular basis. Spread feeding and the use of feeding trays are also feasible solutions. Proteins, carbohydrates, lipids, vitamins, and minerals are all required for development and upkeep and must be supplied by the feed. When storing feed, be certain that the quality does not deteriorate.

Fish farming in cages is less labor-intensive than farming in ponds. Harvests can be either extensive or selective by towing the cages to the best site. Selling fish when they are alive may be a lucrative side business.

The purpose of people in charge of a cage culture facility is to maximize output while spending as little money as possible. To ensure that cultured fish develop at the projected rate in proportion to feeding rate and stocking density, losses due to disease and predators are reduced, environmental factors are monitored, and technical facility efficiency is maintained; all of this is dependent on proper management. Furthermore, constant physical maintenance of cage structures is essential. The netting enclosures should be examined on a regular basis. Rope anchors and net cages require rapid repair and upgrading. A monthly net exchange is also advised since it keeps the net healthy by eliminating garbage and uneaten food and decreasing the negative impacts of fouling.

6. Contamination from cage nets

Several cases of cage farming have revealed signs of fouling of cage nets and other structures. Over time, nets become biofouling. Before the development of molluscan fouling can proceed, more research on the fouling induced by edible oysters and sand barnacles is required. We can eliminate algal mats and other kinds of periphyton more efficiently by introducing omnivorous grazers to enclosures. A fouled net will encounter greater resistance and hence lose more of its catch as a result. A new net may need to be erected every two to four weeks, depending on the frequency of fouling. Oyster fouling, which occurs immediately after the periodic spat fall, is a critical phase for net exchange. Herbivorous fish like rabbit fish, pearl spot, and scat may be used to manage biofoulers, but their wide implementation would need thorough research.

7. Disease monitoring

Changes in behavior, particularly around feeding time, can be utilized as early warning indicators for the state of fish populations, therefore they must be monitored.

2.4. RUNNING WATER AQUACULTURE SYSTEM

Despite the fact that the vast majority of Indonesia's aquaculture takes place in ponds or other types of still bodies of water for the purpose of hatching eggs and raising juvenile or adult fish, there are some farms in remote mountainous areas that make use of moving water. Even if there is a higher demand from customers for seafood, fish that are raised in systems with stagnant water take much longer to reach market size. It is necessary to find a solution to this problem, which is faced by fish farms everywhere.

In the early 1970s, researchers came up with a method that allowed for the fast reproduction of fish in moving water. Running water systems are a type of intensive cultivation that demand frequent water changes, a sizable amount of water, and a high population density of the species that is being cultivated. Aquaculture on flowing water offers a number of benefits to its participants. Not only will there be an enough amount of dissolved oxygen in the water, but it will also be able to remove any waste products or phony food pellets that were not consumed by the animal.

Because running water systems require fish stock to weigh at least 100 grams, the demand for fish produced by fish farmers that weigh 100 grams increases. Additionally, enormous swaths of land are not required, the amount of time it takes for fish to reach a size at which they can be sold is reduced, and fish farmers' revenue increases (Figure 2.4).

Fish that want to survive in the intense currents of a swarming population need to be both hardy and social. Because of the high stocking density, it is essential that the fish have a high level of disease resistance. Last but not least, the fish need to have their appetites trained as quickly as possible to accept pellets and other types of artificial food.

Figure 2.4. *Running water aquaculture system.*

These conditions have been met by the common carp, which has been utilized in a similar aquaculture setting in Indonesia. Because of its great demand and economic potential, this is the most significant freshwater fish to produce. The majority of fish farms are found in West Java, West Sumatra, North Sumatra, and North Sulawesi.

2.5. ZERO WATER EXCHANGE AQUACULTURE SYSTEM

The growing worldwide demand for aquaculture products has accelerated its growth. Aquaculture's fast rise has led to significant challenges such as habitat destruction, ecological degradation, eutrophication, and disruptions to hydrological cycles. One of the most significant concerns confronting aquaculture is fertilizer runoff from agricultural ponds contaminating water sources. The most serious problem in the industry is the spread of diseases caused by a plethora of potentially lethal viruses. Another important concern associated with aquaculture is eutrophication. As a result, the pond's carrying capacity is diminished, putting unnecessary pressure on the aquaculture animals.

Long-term solutions for protecting aquaculture from disease outbreaks and reducing hazardous wastewater discharge are critical. A zero-exchange system may be more appropriate for resolving these challenges than traditional pond-based intensive aquaculture production methods. When generated utilizing the zero water exchange aquaculture production method under normal conditions, plankton feeds used in aquaculture are predicted to be both affordable and nutrient-rich. Aquaculture animals, such as shrimp and fish, may consume natural biota such as phytoplankton, zooplankton, and bacteria. In intensive and semi-intensive agricultural systems, microalgae and other microorganisms play an important role in nutrient recycling, minimizing anoxic zones in ponds, lowering nutrient loads in the effluent, and feeding shrimp. Because of the sun's free energy, outdoor pond algae growth is free or extremely cheap. However, due to contamination problems with predators, parasites, and "weed" algae species, this strategy is only effective for a small number of fast reproducing organisms. Furthermore, outdoor production necessitates a huge amount of appropriate land, which is susceptible to weather fluctuations (Figure 2.5).

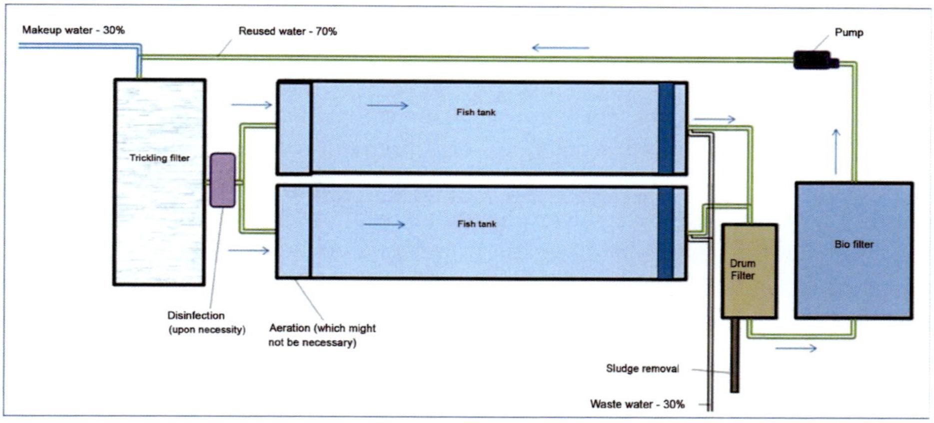

Figure 2.5. *A Recirculating aquaculture system model.*

Aquaculture and its fast expansion can be attributed in large part to several technological advances in the production of phytoplankton and/or zooplankton. Artemia and other phyto- and/or zooplankton species, both of which are widely utilized as initial meals, have been credited with success. The increasing need for plankton feed as aquaculture has delayed the rate of industrialization of larviculture. Algal cultures may be cultivated inside as long as the cultivator monitors for symptoms of contamination. However, the expense is more than for outdoor culture since more room, power, and professional personnel are required. Biomass production is highest under heterotrophic settings rather than photoautotrophic ones. You can develop a steady aquaculture production system if you can maintain a consistent output of high-quality plankton. This is one of the most important variables for the long-term growth of aquaculture. Plankton culture is critical for the growth of all cultivable secondary and tertiary producers since they form the foundation of the aquaculture food chain. Many farmed animals tend to require phytoplankton and zooplankton in their meals at some point in their development, so this is not surprising. Because of the importance of plankton in aquaculture, several fish, bivalve mollusk, and crab firms have made significant investments in plankton production. Because they do not match industrial criteria, plankton generation technologies are frequently inefficient and e□pensive. The cost of production, contamination control, nutritional value, and quality control of plankton are all critical components in plankton farming. By employing phyto- and zoo-plankton as food, Indian aquaculture enterprises might use a number of cutting-edge technologies to increase plankton culture productivity and boost product returns per unit area. Without effective bio-remediation, changes in water quality, toxicity, a reduction in soil redox potential, and a subsequent decrease in plankton production may occur in zero water exchange systems.

Aquaculture without water exchange frequently affects the number of growing animals by degrading the quality of the water and sediment. Implementing techniques for zero-water-exchange aquaculture systems, in addition to suitable measures such as pond preparation and maintenance, and bioremediation technology, can help to preserve sustainability and economic viability. The use of zero-water-exchange aquaculture systems for plankton is one approach for better managing environmental concerns and preventing the spread of infectious illnesses. It aids in the monitoring of effluent quality and reduces the need for cultural space. A lot of food and oxygen must enter the system for it to produce a thick culture. Waste management is critical because improper waste management can generate stress and impede the system's capacity to expand and operate. In bigger aquaculture ponds, leftover feed and garbage are frequently digested naturally by a diverse range of microorganisms or utilized by both phyto- and zooplankton in integrated aquaculture systems. A recirculation system can solve a number of difficulties with the culture system.

Aquaculture currently incorporates a wide range of tools and tactics to maximize productivity. Addressing the multiple environmental issues in a high-density aquaculture setting is more cost-effective by using a recirculating system in which untreated wastewater is treated outside the system. RAS cultivates and gathers phytoplankton in addition to biological filters or integrated biofloc systems to remove undesirable elements such as nitrogen and phosphorus from wastewater. Aquaculture systems, since they are self-contained, integrated aquaculture systems, help to maintain water quality. In multi-trophic systems, feeding garbage to aquatic plants, algae, and zooplankton provides "natural" conditions. A variety of factors, including design, can influence the profitability of this sort of multi-trophic aquaculture. A multi-trophic recirculating aquaculture system for fish and other species has been designed to ensure the long-term viability of aquaculture, contribute to environmental security, and supply food for the world's rapidly growing population. Because of the availability of suitable food-chain-like organisms, farmers may make more money by collecting and selling phyto- and zooplankton-related byproducts. Zero Water Exchange Aquaculture Pond System has recently acquired favor in aquaculture due to its excellence in disease outbreak and environmental pollution management.

In systems with restricted outflow, phytoplankton and microbial populations play a significant role in the recycling of natural nutrients. The number of microbes, plants, and animals in a particular system determines carrying capacity primarily. One of the most compelling aspects of zero-water-exchange culture technology is its ability to increase biosecurity while lowering feed costs and water use. As a result, aquaculture has a higher chance of being more environmentally friendly in the future. Technological improvements that increase productivity and farmer income while reducing environmental problems are critical to the long-term viability of the aquaculture business. Aquaculture may now progress with improved disease prevention and environmental impact.

REFERENCES

1. Abdel-Hakim, N. F., Bakeer, M. N., & Soltan, M. A. (1999). Integrated fish culture with farm animals. *Annals of Agricultural Science, Moshtohor, 37*(2), 1001–1015.
2. Abdel-Hakim, N. F., Bakeer, M. N., & Soltan, M. A. (2000). Effect of two manuring systems on water quality and plankton communities in fish ponds. In *Conference of Social and Agricultural Development of Sinai, El-Arish-North Sinai, 16–19 May, Egypt* (pp. 147–158).
3. Abdel-Hakim, N. F., Bakeer, M. N., & Soltan, M. A. (2000). Fish culture in rice fields at Kafr El-Sheikh. In *The First International Conference (Agricultural Economics in Islamic World)*, Al Azhar University, Cairo, Egypt.
4. Abdel-Hakim, N. F., Bakeer, M. N., & Soltan, M. A. (2000). Growth performance of Eel (*Anguilla anguilla*), Nile tilapia (*Oreochromis niloticus*), and grey mullet (*Mugil cephalus*) cultured in cages under two feeding systems. In *Proceedings of the 3rd All Africa Conference on Animal Agriculture & 11th Conference of the Egyptian Society for Animal Production*, Alexandria, Egypt, (pp. 329–335).
5. Abdel-Hakim, N. F., Bakeer, M. N., & Soltan, M. A. (2000). Technical and economic feasibility of integrated fish-duck culture in Egypt. In *The First International Conference (Agricultural Economics in Islamic world)*, Al Azhar University, Cairo, Egypt, 22–25.
6. Abdel-Hakim, N. F., Bakeer, M. N., & Soltan, M. A. (2001). Effect of dietary protein levels on growth performance and pond productivity of Nile tilapia (*Oreochromis niloticus*), Eel (*Anguilla anguilla*) and grey mullet (*Mugil cephalus*) reared in polyculture system. *Egyptian Journal of Aquatic Biology and Fisheries, 5*(4), 61–85.
7. Abdel-Hakim, N. F., Hussein, M. S., Bakeer, M. N., & Soltan, M. A. (2001). Effect of protein level and stocking density on growth performance of Nile tilapia (*Oreochromis niloticus*) cultured in tanks. *Egyptian Journal of Nutrition and Feeds, 4*, 763–780.
8. Abdel-Hakim, N. F., Soltan, M. A., & Bakeer, M. N. (2000). The technical and economic feasibility of cultivating Nile tilapia (*Oreochromis niloticus*) in rice fields under the Egyptian conditions. In *Conference of Social and Agricultural Development of Sinai, El-Arish-North Sinai, 16–19 May, Egypt* (pp. 171–182).
9. Bakeer, M. N., Soltan, M. A., Tharwat, A. A., & Samra, I. M. (2005). Studies on over-wintering of Nile tilapia (*Oreochromis niloticus*) fry. *Annals of Agricultural Science, Moshtohor, 43*(3), 1067–1082.

10. Beveridge, M. C. M. (1996). *Cage Culture*. Fishing News Books.
11. Environmental Protection Agency. (1973). *Wastewater Treatment and Reuse by Land Application*. Cincinnati, Ohio: Environmental Protection Agency.
12. Franks, F. (2000). *Water*. Royal Society of Chemistry.
13. Gleick, P. H. (1996). Water resources. In *Encyclopedia of Climate and Weather* (Vol. 2, pp. 817–823). Oxford University Press.
14. Ibrahim, M. K., Hafez, F. A., & Soltan, M. A. (2000). Effect of organic fertilization, supplementary feeding, and stocking rate on growth performance of Nile tilapia and silver carp. *Egyptian Journal of Agricultural Research, 78*(4), 1775–1799.
15. Soltan, M. A. (2002). Using of tomato and potato by-products as non-conventional ingredients in Nile tilapia (*Oreochromis niloticus*) diets. *Annals of Agricultural Science, Moshtohor, 40*(4), 2081–2096.
16. Soltan, M. A., Abdel-Hakim, N. F., & Bakeer, M. N. (1999). Effect of stocking rate, organic fertilization and supplementary feed on growth performance, carcass and chemical analysis of Nile tilapia (*Oreochromis niloticus*). *Egyptian Journal of Nutrition and Feeds, 2*, 765–778.
17. Soltan, M. A., Abdella, M. M., El-Sayaad, G. A., & Abu-El Wafa, M. H. (2015). Effect of some over-wintering regimes on survival and growth performance of Nile tilapia (*Oreochromis niloticus*). *Biological and Chemical Research, 2*, 362–374.
18. Soltan, M. A., Bakeer, M. N., & Samra, I. M. (2006). Effect of some feeding regimes on water quality, growth, and productivity of Nile tilapia (*Oreochromis niloticus*) reared in earthen ponds. *Egyptian Journal of Agricultural Research, 84*(1A), 1–18.
19. Soltan, M. A., Radwan, A. A., & Samra, I. M. (2002). Effect of varying protein, energy and protein to energy ratio on growth, feed efficiency and body composition of Nile tilapia (*Oreochromis niloticus*). In *1st Annual Conference of the Egyptian Aquaculture Society*, Al-Arish, North Sinai, Egypt, 13–15.

Chapter 3

Intelligent Fish Farms

Contents

INTRODUCTION

With the continued expansion of aquaculture scale and density, modern aquaculture methods have been forced to overproduce, resulting in an accelerated imbalance rate of the water environment, the occurrence of fish diseases, and a decline in aquatic product quality. Furthermore, because the average age of agricultural workers in many parts of the world is increasing, fishery production will face a labor shortage, and aquaculture methods must be changed immediately. Modern information technology has gradually penetrated various fields of agriculture, and the concept of an intelligent fish farm has begun to emerge. The intelligent fish farm attempts to deal with the precise work of increasing oxygen, optimizing feeding, reducing disease incidences, and accurately harvesting using the concept of "replacing human with machine," in order to completely liberate manpower and realize green and sustainable aquaculture. This chapter examines the use of fishery intelligent equipment, IoT, edge computing, 5G, and artificial intelligence algorithms in modern aquaculture, as well as current issues and future development opportunities. Meanwhile, based on various business requirements, design frameworks for key functional modules in the construction of an intelligent fish farm are proposed.

3.1. CONCEPT OF INTELLIGENT FISH FARMS

Fishery products are crucial in ensuring food security and nutrition strategies at all levels. The global fish production in 2018 is estimated to be approximately 179 million tons, with aquaculture contributing 82 million tons. Between 1961 and 2017, the global consumption of fish for food experienced an average annual growth rate of 3.1%, which was nearly double the average annual growth rate of the world population during the same period. This growth rate also surpassed that of other animal protein foods, such as meat, dairy products, and milk, which had an annual growth rate of 2.1% (FAO 2020). Fish possess significant potential to serve as a valuable alternative source of protein in lieu of livestock and poultry. The primary factor for this is the presence of inexpensive fish protein of superior quality (Gopalakrishnan et al., 2017), coupled with the absence of widespread issues such as avian influenza in aquaculture. The global fishing industry has faced challenges in obtaining migrant labor due to the travel and entry-exit restrictions imposed as a result of the COVID-19 pandemic. In addition, the agricultural labor force involved in fishery production is progressively reaching an advanced age, and a scarcity of labor is impeding the ability to ensure the consistent provision of aquatic products worldwide. The deterioration of aquaculture water quality is primarily caused by overfishing, outdated aquaculture methods, human activities, industrial pollution, and other factors (Sarma et al., 2013; Peters and Meybeck 2000). As a result, aquatic ecosystems are harmed, and there is an increased occurrence of diseases in aquatic products. Many significant aquaculture production regions worldwide are encountering challenges due to increasingly strict domestic environmental protection policies and worsening aquatic biological diseases (Vatsos and Angelidis, 2017). Concurrently, the appropriate areas for aquaculture will gradually decrease, leading to a decline in the profitability per unit of land. Insufficient scientific and technological innovation and engineering technical support exist throughout the entire fishery industry chain. Insufficient fundamental research and theoretical understanding exist regarding the interaction between biological environments and the integration of engineering processes. The progress in aquaculture engineering, mechanization, information technology, and equipment construction is significantly inadequate. Breeding enterprises in underdeveloped areas face limitations due to inadequate breeding technology and unfavorable environmental conditions. The breeding germplasm is deteriorating, and enterprises suffer from a lack of high-quality varieties, resulting in low-quality aquatic products (Brugère et al., 2019). In addition, the utilization of diverse antibiotics, hormones, and high-residue chemical drugs in aquaculture production is no longer an isolated occurrence (Mateo-Sagasta et al., 2017). The improper and unscientific administration of drugs has brought the issue of drug residues in aquatic products to the forefront of concern.

Aquaculture is encountering significant obstacles, yet it also presents a substantial prospect. The future development directions of aquaculture include ecological, facility, industrial, and intelligent aspects (Francesca et al., 2020; Martin et al., 2018). ABB, a global leader in automation technology, is constructing the first innovative offshore submersible fish farm in the Arctic Ocean to enhance Norway's salmon production and uphold stringent quality standards (ABB 2019). The fish pens will be operated through remote control using a feed barge. Automated equipment collects environmental data, such as meteorological conditions, ocean currents, dissolved oxygen (DO) levels, water temperature, pH values of various water layers, and net cage biomass. This data is then transmitted wirelessly and uploaded to ABB Ability™. The ballast water system in fish pens is designed to withstand severe weather and wave conditions. The industrial electrical system possesses a high level of durability, but it does not play a role in the analysis of fish behavior, intelligent feeding, or the harvesting of adult fish. Several organizations have initiated the construction of diverse forms of intelligent fish farms. The ARTIFEX project aims to conduct inspection, maintenance, and repair operations at fish farms located at sea. This is achieved through the collaboration of an unmanned ship, an unmanned aerial vehicle, and a remotely controlled underwater robot (Artifex, 2018). Aside from serving as a launch and landing site for airborne drones, the unmanned vessel can also function as a central station for subaquatic robots. The unmanned aerial vehicle is tasked with inspecting the structural soundness of surface facilities and overseeing the feeding process of fish. The underwater robot is required to not only perform the underwater inspection task, but also undertake the cleaning and repair of the fishing net. These sophisticated devices will offer technical assistance for the swift implementation of an advanced fish farming facility. The main technologies utilized in the aforementioned projects encompass the hardware design of unmanned ships, aircraft, and underwater vehicles, as well as the software development of autonomous control systems. The challenge of these projects lies in the requirement for seamless integration and precise coordination of these sophisticated technologies. Intelligent fish farms, as a highly developed form of aquaculture, utilize intelligent machines to replace manual labor and optimize the allocation of fishery resources in a data-driven manner, resulting in efficient, environmentally friendly, and intelligent aquaculture.

This chapter presents the most recent integration of edge computing and 5G technology in the development of an intelligent fish farm. Additionally, it introduces the architecture of the intelligent fish farm ecosystem for the first time. Ne□t, the key intelligent equipment of the intelligent fish farm is listed based on the system's hardware configuration, and its technical characteristics are analyzed. The paper provides a comprehensive overview of the current research on the top-level design of intelligent fish farms. It covers various aspects such as water quality monitoring and control, intelligent feeding, fish behavior monitoring, biomass estimation, fish disease diagnosis, and equipment fault diagnosis. Additionally, the paper proposes a design scheme for the relevant business modules of an intelligent fish farm. This chapter concludes the challenges and areas of current research in the development of intelligent fish farms. Its aim is to offer aquaculture professionals up-to-date, comprehensive, and informative ideas for future fishery construction.

3.1.1. Definition and System Framework of Intelligent Fish Farm

An intelligent fish farm is a fully automated production system that operates in all weather conditions and covers the entire production process and space. This means that workers do not need to enter the fish farm as advanced technologies such as IoT, big data, artificial intelligence (AI), 5G, cloud computing, and robots are utilized for remote measurement and control of the fish farm. These technologies enable independent control of fishery facilities, equipment, and machinery by robots, allowing for the completion of all production and management operations of the fish farm. Intelligent fish farms utilize digital and intelligent technology to address the challenges of labor shortage, water pollution, high risk, and low efficiency in aquaculture. An intelligent fish farm refers to the industrial transformation of fishery production methods and represents the future direction of the fishery industry. The classification of intelligent fish farms is based on the different cultural environments and includes four categories: pond-type, land-based factory-type, cage-type, and intelligent marine ranch.

Aquaculture facility with a pond design utilizes sensors to gather real-time data on water quality, while unmanned aerial vehicles monitor fish activity on the water surface. Bionic fish are used to monitor the growth status and feeding process of fish. The water quality is controlled through the application of fertilizers and the use of chemical sprays on an unmanned boat. The transportation of bait is carried out by an unmanned vehicle. The intelligent aeration system precisely controls dissolved oxygen (DO). The sophisticated bait feeder achieves precise and automated feeding. Automated trawl machines and fish distributors are used for intelligent harvesting. The reliable operations of the pond-type intelligent fish farm can only be guaranteed when all systems function cohesively and in an organized manner.

A land-based factory-style intelligent fish farm primarily achieves automated recirculating aquaculture (RAS). The fish farm utilizes micro-filters, biological filters, intelligent feeding machines, aquaculture tail water purification and utilization devices, as well as advanced model and intelligent equipment technologies to create a fish circulation three-dimensional aquaculture model. Through an extensive analysis of the fundamental needs of aquaculture biology and the operational parameters of the RAS system, a scientific approach is taken to determine the optimal culture density, suitable water environment requirements, and effective aquaculture management strategy for benthic fish in the recycling aquaculture mode. This is achieved by gathering production data and utilizing advanced big data analysis technology. The fish farm implements advanced seed production and selection technologies to establish a comprehensive breeding system for circulating aquaculture. This system covers all stages of parent fish mating, egg hatching, fry cultivation, adult fish breeding, sales, and packaging.

A cage-based automated fish farm utilizes sensors to gather data on seawater quality and ocean currents, while machine vision and sonar are employed to collect information on fish movement and feeding behavior. The feed barge has the capability to precisely distribute bait based on factors such as water quality, fish biomass, and feeding behavior. The net washing robot autonomously cleans the underwater net. The process of automatically catching fish is achieved through the utilization of an automated net lifting system and a fish pump. The emergency management of water quality is achieved through the use of light supplementation and an emergency o□ygen system. The fish farm is categorized into two types: floating and fi□ed. The aquaculture system will be organized into a three-dimensional structure based on the specific water layers where different aquaculture organisms reside. The bait, energy, and other production materials will be conveyed from the onshore warehouse to the breeding area by unmanned vessels. An intelligent fish farm in the form of a cage can alleviate the strain on offshore resources and efficiently address the issue of land-based aquaculture occupying significant amounts of space.

Smart marine ranches typically employ high-resolution surface cameras and underwater robots to capture real-time video footage of the ranch. This video is then transmitted to a data server in a shore-based information control center via a transmission network. The purpose of this process is to facilitate biological identification, behavior analysis, and biomass estimation. The ranch acquires ocean current data and the distribution of plants and animals near artificial reefs by installing Doppler sensors, underwater cameras, and sonars. The marine ranch employs water quality monitoring buoys, weather monitoring stations, management boats, radar, and other underwater mobile monitoring equipment to achieve comprehensive automated monitoring of the marine environment. Satellite remote sensing technology (SRS) allows marine ranches to assess the fluctuations in fish stock across various spatial and temporal scales (George 2014). Synthetic aperture radar remote sensing technology can be used to remotely detect ocean currents, cyclones, and other marine environmental information related to fisheries by analyzing the interaction between microwaves and the micro-scale structure of the sea surface. Ranchers have the option to utilize unmanned aerial vehicles to distribute bait in the air or unmanned boats to autonomously feed by approaching the water surface. The transportation of material supply for marine ranches typically involves the use of unmanned boats in a land-sea relay fashion.

Figure 3.1. displays the key technologies utilized in intelligent fish farming. The four types of intelligent fish farms all consist of specific fishery infrastructure, as well as an above and underwater environment monitoring system, a water quality and feeding control system, an unmanned vehicle platform for auxiliary monitoring, control, and transportation tasks, and an adult fish harvesting system. In the future, the intelligent fish farm will have the capability to choose the appropriate hardware platform based on specific aquaculture needs. The collaboration among devices is facilitated through the utilization of the Internet of Things (IoT) and 5G technology. The accurate calculation of intelligent algorithms is crucial for achieving precise control of equipment.

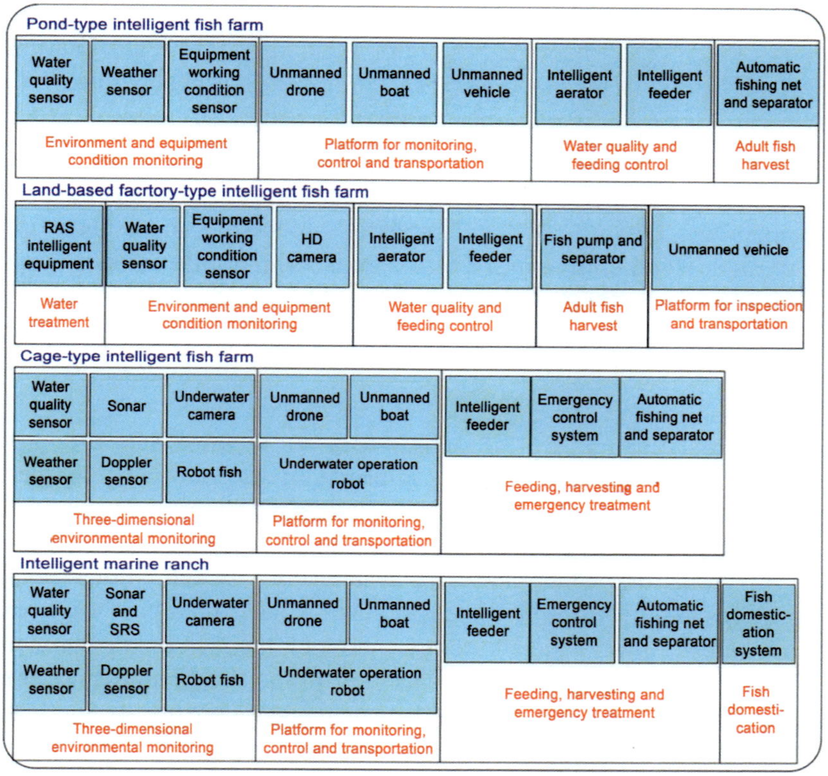

Figure 3.1. *The involved key technologies in intelligent fish farm.*

3.1.2. Advance Information Technology in Intelligent Fish Farm

The conventional aquaculture IoT system is a highly integrated combination of cloud technology and Internet of Things technology (Huan et al., 2020). The system is organized into three tiers: the device layer, the network layer, and the cloud service layer. The device layer consists of sensing equipment, control equipment, and data acquisition terminals. The sensing equipment is tasked with gathering environmental data, including dissolved oxygen (DO), pH, temperature, salinity, ammonia nitrogen, nitrite, water level, etc., as well as monitoring the operational status of the device and capturing video images of aquaculture activities. The control equipment comprises an aerator (oxygen cone), feeder, pump valve, and other aquaculture equipment. The data acquisition terminal is responsible for transmitting sensor data and receiving control instructions. The network layer typically utilizes wireless network technologies, such as Bluetooth, Wi-Fi, 3G/4G, Lo-Ra, NB-IoT, and others, to facilitate data transmission between the device layer and cloud service layer. The cloud service layer encompasses both the cloud platform and smartphone app, primarily utilized for the storage and processing of aquaculture data, as well as the provision of diverse aquaculture information services.

Despite the benefits of easy equipment access and quick system setup, the aquaculture IoT system faces challenges in handling the increasing number of equipment and data due to the centralized processing and unified upload of data through cloud computing. Sensors consistently gather diverse sensing data, which typically remains stable or e□periences minimal fluctuations. Transferring all the data to the cloud for processing will require a significant amount of network and cloud resources. Insufficient network stability hinders the timely processing of data and control of equipment.

It is necessary to transmit all sensor data and control data through the network, but there are potential risks of information eavesdropping, tampering, deception, and unauthorized operation of equipment. These issues will result in higher e□penses for the IoT system (such as communication flow, information storage, and computing costs), decrease system stability and availability, and pose challenges in achieving automated production, particularly in the context of large-scale industrialized aquaculture. Furthermore, the lack of standardization in sensing and control devices, as well as their limited computing and storage capabilities, have posed significant challenges to equipment access and linkage control. Hence, in order to establish an advanced aquaculture facility, it is imperative to thoroughly investigate and address the aforementioned issues.

Edge computing enables the provision of intelligent services at the periphery of the network, in close proximity to the source of data or devices. This allows each edge of the Internet of Things (IoT) to possess capabilities for data collection, analysis, computation, communication, and intelligent processing. Consequently, data can be processed, filtered, and analyzed in close proximity to its origin (O’Grady et al., 2019). Local decision-making and event processing can fulfill the essential criteria of network capacity constraints, timely data processing, limited resources, and addressing security and privacy concerns. In the past, the application of machine learning and deep learning was limited to high-performance hardware, which involved edge training and inference carried out by gateways, edge servers, or data centers. Currently, there is a significant amount of interest and investment in the field of research and development focused on microcontrollers that are capable of performing machine learning tasks at the edge.

This area of study is commonly referred to as TinyML (Tiny Machine Learning). Current water quality sensors lack robust self-verification mechanisms to confirm their operational status. Consequently, it has been assumed that these sensors are functioning properly without any concrete evidence. When the sensor malfunctions, the resulting output will significantly diverge from the real situation, potentially leading to false alarms and diminishing the reliability of the detection outcomes. Zhao et al., (2011) initially utilized edge computing to address the issue of sensor fault diagnosis and data repair.

Multiple classifiers derived from a binary classifier using relevance vector machine are utilized to perform fault diagnosis on pH, temperature, and salinity sensors. The relative nature of multiple parameters in water quality monitoring is effectively utilized to achieve data recovery of a multi-function self-validating sensor, even when there are only a few nonlinear training sets available.

As the 5G network era begins, intelligent fish farms also demand higher network performance. The capability index of the 5G network has been greatly improved compared to the 4G network. The maximum allowable delay is 1 millisecond, the peak data rate can reach 10 gigabits per second, the connection density can reach 10 terminals per kilometer, and the user experience rate can reach 100 megabits per second.

The existing network architecture faces challenges in meeting these requirements. Currently, a large volume of data from Internet of Things (IoT) devices is being stored in cloud service centers. The real-time communication between users and the cloud center necessitates a high transmission rate, which can lead to significant network congestion in areas with high mobility. Businesses that have strict real-time demands require end-to-end low latency at the millisecond level. However, conventional cloud computing models struggle to fulfill these requirements.

Hence, this chapter presents the integration of Edge computing and 5G technology into the aquaculture Internet of Things (IoT) system with the aim of enhancing the system's standardization, stability, and usability. Figure 3.2. displays the conceptual design of an intelligent fish farm. Distributed systems and cloud services are utilized to distribute resources, while crucial applications and fundamental process control should be implemented at the edge layer. Network/cloud deployment is possible for web services and data analysis applications. The communication layer and the edge layer combine resources and ensure their compatibility.

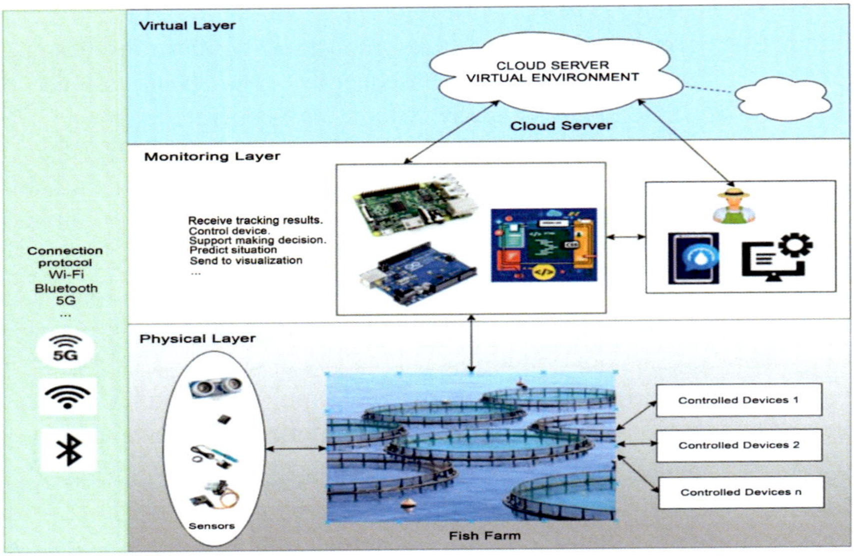

Figure 3.2. *Conceptual design of IoT ecosystem in intelligent fish farm.*

The implementation of fishery informatization will generate a substantial volume of multidimensional data. Big data technology refers to the rapid processing and comprehensive analysis of large volumes of data in order to extract valuable insights. The process encompasses the acquisition, preprocessing, storage, and analysis of large volumes of data. AI can utilize its ability to simulate human cognition and intelligent behavior to process the vast amount of information from the IoT and big data. It can analyze and evaluate problems, ultimately making decisions and ensuring precise operations in the fish farm. In contrast to conventional technology, AI technology emphasizes the computation, processing, analysis, prediction, and planning of problems. The Internet of Things (IoT), edge computing, 5G, Big data, and AI work together and integrate closely to offer technical assistance for the development of intelligent fish farms.

3.2. INTELLIGENT EQUIPMENT AND ROBOTS IN INTELLIGENT FISH FARM

Traditional aquaculture primarily relies on anecdotal knowledge with limited effectiveness, significant expenses, and minimal mechanization. Currently, various crews on service vessels are responsible for performing routine tasks on fish farm facilities, including monitoring fish welfare, inspecting facilities, controlling feed rationing, and counting lice (Baird 2017). In the future, autonomous and remotely operated systems could have a significant impact on performing various tasks at fish farm facilities. This is because the working conditions in aquaculture, especially in marine cage culture, are complex and constantly changing. Intelligent equipment and robots will incorporate sophisticated sensors, large-scale data analysis, and artificial intelligence (AI) to effectively adjust to intricate operational settings and achieve autonomous operation in intelligent fish farms. In addition, they have the potential to significantly decrease the amount of labor required and the associated expenses, while enhancing the efficiency of fishery production. Furthermore, advanced machinery and robots have the capability to independently perceive the extensive data from intelligent fish farms and transmit this data to the cloud platform, serving as the foundation for big data analysis (Mustapha et al., 2021).

AI is used to enhance the capabilities of intelligent equipment and robots in fishery production by providing them with the ability to learn, analyze, and make independent decisions. Robot technology is the synergistic integration of information technology, machinery, and automation technology. According to Rembold (2020), it has the capability to allow various types of machines to actively engage in every stage of fishery production, similar to how humans participate. Furthermore, the successful implementation of intelligent equipment and robots necessitates the utilization of edge computing, machine vision, navigation, and precise control technology. This is crucial in order to fulfill the requirement of machines replacing humans, as stated by Shi et al., (2016).

Intelligent equipment in the field of fishery utilizes modern information technology and intelligent manufacturing technology to undergo digital transformation. This enables precise operations in intelligent fish farming. Farm intelligent equipment and robots can be categorized into mobile and stationary equipment. Mobile equipment refers to equipment that requires movement to achieve autonomous operation. This category primarily includes unmanned ground vehicles (UGVs), unmanned aerial vehicles (UAVs), unmanned ships or surface vessels (USVs), and unoccupied underwater robots (ROVs). The fi□ed equipment comprises water quality monitoring and control devices, feeding devices, o□ygen devices, as well as harvesting devices like fish suction pumps and classification equipment.

3.2.1. Intelligent Hardware for Measure and Control

Water environment ecological monitoring involves the utilization of sensors and cameras deployed on unmanned vessels or surface buoys to autonomously gather data on water quality parameters such as dissolved oxygen, temperature, chlorophyll, turbidity, ammonia nitrogen, and pH. This monitoring also includes capturing images and videos of aquatic organisms in aquaculture settings. The collected data is then stored, transmitted, analyzed, and used for predictive purposes. Consistently precise detection of water quality parameters in aquaculture allows for a dependable data source to enable automatic control and intelligent decision-making in smart fish farms. Currently, the detection of dissolved oxygen (DO) primarily involves the utilization of the Clark electrode method and the fluorescence quenching method (Tai et al., 2016). Optical dissolved oxygen sensors do not require calibration and offer several benefits, including rapid response time, consistent measurement accuracy, robust resistance to interference, and minimal maintenance requirements. Future research will primarily focus on developing membrane-free DO sensors that utilize fluorescence and nanocomposite modified electrodes. The online measurement of pH in aquaculture is primarily achieved using digital differential electrodes or composite electrodes (Lee et al., 2017). The digital differential electrode employs a three-electrode system rather than the conventional double-electrode technology, resulting in superior accuracy and reliability. In aquaculture, the detection of ammonia nitrogen, nitrate nitrogen, and nitrite nitrogen typically relies on online water quality analyzers. These analyzers transfer the entire laboratory chemical analysis process to a detection box located on the shore. The automatic analyzer consists of a water sample pretreatment module, a chemical digester, a titrator, an optical sample pool, a peristaltic pump, and an electronic detection unit. The equipment is capable of supporting linkage control, fully automated operation, maintenance-free functionality, and can achieve automatic zero adjustment, automatic calibration, and automatic cleaning. The analysis results can be transmitted to the data center either through fieldbus or wireless communication. In the coming years, unmanned boats and underwater robots with aquaculture environment dynamic monitoring systems will be able to autonomously navigate intelligent fish farms. They will also be capable of conducting comprehensive, real-time, and multi-location water quality monitoring tasks. This advancement will significantly enhance the level of informatization, automation, and intelligence in intelligent fish farming.

An intelligent aeration system is a set of equipment that is capable of precisely measuring and controlling the dissolved oxygen (DO) levels in water. It consists of multiple sensors, a network transmission module, and an IoT actuator (Huan et al., 2020). The sophisticated aerator has the capability to continuously monitor real-time water temperature, air humidity, air pressure, and dissolved oxygen (DO) levels. Simultaneously, it can capture scene data through video surveillance and transmit this data to the cloud platform, enabling accurate control of the aerator. Furthermore, the intelligent aerator has the capability to accurately forecast the dissolved oxygen (DO) levels within a brief period. A recirculating aquaculture system (RAS) is capable of reusing the wastewater generated in an aquaculture pond by subjecting it to a sequence of water treatment units (Xiao et al., 2018). RAS incorporates cutting-edge technologies, including environmental engineering, civil engineering, modern biology, nano engineering (Moges et al., 2019), and electronic information, to eliminate detrimental contaminants such as feces, ammonia nitrogen, and nitrite nitrogen from aquaculture water. The system primarily comprises a protein separator, microfiltration machine, biofiltration system, degassing system, temperature control system, sterilization system, water quality monitoring system, and oxygenation system. The primary concerns in the design of commercial RAS are the swift elimination of ammonia nitrogen and the rapid elevation of dissolved oxygen levels. Researchers (Pedersen et al., 2020; Wik et al., 2009) have utilized the theory of material balance to optimize the design parameters of the system. This includes the amount of water supply, oxygen supply, circulation amount, circulation times, and effective volume of the biofilter. They achieved this by dynamically simulating the process of fish growth, waste generation, and water treatment cycle. Computer-aided dynamic simulation will prove to be a highly efficient tool for the design of RAS. A specialized data collector will gather the operational parameters of the motor, including the working current, working voltage, vibration amplitude, and surface temperature. This will enable real-time monitoring of the circulating pump's operation status. This data will be used as evidence for diagnosing faults in the aquaculture water treatment system.

3.2.2. Intelligent Hardware for Precision Self-Feeding

The automatic feeding system has found extensive application in industrial recirculating aquaculture. This includes the use of both the automatic feeding system with multi monomer centralized control and the automatic feeding robot system. The implementation of automatic feeding systems in developed countries like Norway, Japan, and the United States has reached the application stage, successfully achieving precise control in the processes of feed transportation, storage, and delivery. The net cage automatic feeding system, developed by a Norwegian fishery equipment enterprise, comprises a management system, an online monitoring system, and a feeding module. The management system has direct control over the fan and feeder, allowing it to make real-time adjustments to the feeding process. The online monitoring system simultaneously monitors multiple parameters of aquaculture water, including pH, dissolved oxygen (DO), and temperature, in real-time. It then transmits the collected data to the management system, enabling the automated feeding of bait (Wang et al., 2021). The robot feeding control system, created by Arvo-Tec, a Finnish company, enables remote control of feeding, enhancement of water quality, and precise feeding through a web interface (Arvotec, 2021). The Arvo-Tec control unit WOLF is a comprehensive system that combines feeding, measurement, light control (photo period), and alarm functions into a single integrated unit. The system utilizes water temperature, oxygen content, biomass, and other environmental parameters as the initial data to regulate feeding control based on an energy demand model. Furthermore, the system calculates fish growth by taking into account the feed conversion ratio. Within the intelligent fish farm that operates in a pond, it is necessary to utilize advanced equipment capable of dropping bait in a smart manner. This equipment should be installed on unmanned boats or unmanned aerial vehicles (UAVs). The UAVs will be responsible for autonomously transporting and loading the bait.

3.2.3. Unmanned Patrol System

The intelligent fish farm's patrol equipment consists of UAVs, USVs, and robot fish. The patrol equipment is capable of performing daily underwater management, video information collection, behavior monitoring of breeding objects, and ecological environment monitoring in aquaculture. The UAV's flight attitude control module utilizes PID and fuzzy control algorithms to achieve hovering, vertical motion, rotation, and pitch (Bonadies et al., 2018). Using GNSS, machine vision, and 5G, the automatic navigation system can determine the patrol UAV's speed, course, height, position, and distance, as well as perform operation path planning and intelligent obstacle avoidance, ensuring high precision positioning and navigation of the patrol UAV in the complex environment of an intelligent fish farm. The monitoring system includes cameras, hyperspectral sensors, and remote sensing equipment that uses machine vision to achieve omnidirectional perception of environmental ecological information and fish feeding information in an intelligent fish farm. 5G will be used in Open-Pen Sea Cage Aquaculture for high-throughput HD video transmission and monitoring, which will effectively solve the problem of high cost and difficulty in laying optical cables. In the future, UAVs using AI algorithms will be outfitted with powerful airborne edge computing modules to enable autonomous takeoff, landing, cruise, and monitoring. Furthermore, the UAV used in the inspection of intelligent fish farms will be equipped with an automatic charging controller module to enable true autonomous operation.

Unmanned vessels can plan their speed, course, position, and operation paths using a GPS or BeiDou navigation system. Unmanned boats can carry underwater monitoring sonar to track fish stocks and estimate biomass. It also completes tasks by interacting with ROV. Unmanned boats can also be used to transport other aquaculture equipment like aerators, bait feeders, and harvesting equipment, as well as to launch UAVs and biomimetic robot fish. Biomimetic robot fish can automatically monitor and inspect underwater aquaculture environments (Ravalli et al., 2017). The underwater acoustic system includes sonar and multi-beam detection equipment capable of underwater positioning, automatic obstacle avoidance, object tracking, and biomass estimation. The biomimetic robot fish carries a variety of sensors that automatically monitor water quality and the status of key equipment. Biomimetic robot fish can also monitor fish feeding patterns using computer vision technology and analyze relevant data to provide a foundation for optimizing feeding strategies.

The inspection robot, which uses deep learning, computer vision, and positioning technology, can detect the location of sick and dead fish and pick them up using an automatic manipulator in conjunction with an optical and acoustic system. Underwater robots in cage culture can also detect damaged and contaminated net clothes and use tools to clean and repair them. The orbital robot can inspect the circulating water pipe network, oxygenation equipment, and feeding equipment in the recirculating aquaculture workshop using a predetermined inspection route. In the future, the cloud platform and intelligent equipment will complete all inspections in intelligent fish farms. USV, UAV, and ROV will collaborate to conduct three-dimensional monitoring of the water, land, and air environments.

3.2.4. Intelligent Harvesting System

The intelligent harvesting system is the final module for fish farms to complete the breeding cycle. Using this system, breeding objects will enter the market via transportation, with or without water. Currently, trawling is the most efficient method of fishing. To achieve precision fishing, the intelligent fish farm system employs sonar, underwater cameras, and a net positioning instrument. The use of unmanned vehicles and ships to transport aquatic products will significantly improve fishery production efficiency. This will help to achieve the goals of high efficiency and energy savings, as well as reducing fishermen's labor intensity. The sea cucumber catching robot uses Doppler-GPS integrated navigation, computer vision, and mechanical servo-drive technology to automatically pick sea cucumbers. The navigation system of the sea cucumber catching robot is in charge of gathering position information from the underwater robot, planning the navigation route, and controlling the robot's movement. The machine vision system collects and preprocesses images of sea cucumbers, recognizes them, and calculates their relative coordinates (Qiao et al., 2017). The actuator can control the pressure pump and manipulator used to pick sea cucumbers.

Intelligent equipment and robots are the key to freeing up aquaculture labor and automating production. They rely on modern information technology, such as comprehensive perception, intelligent processing, intelligent navigation, and automatic control technology, as well as improvements to traditional aquaculture equipment, to achieve unmanned production, environmental information monitoring, optimization control, and precise operation. Robots, unmanned vehicles, unmanned boats, and robotic fish will all play important roles in the intelligent fish farm.

3.3. DATA-DRIVEN DECISION-MAKING

The primary goal of intelligent aquaculture is to move beyond data to decision. To achieve intelligent aquaculture, all aspects of aquaculture must be fine-tuned, including intelligent feeding, water quality control, behavior analysis, biomass estimation, fish disease diagnosis, equipment working condition monitoring, and fault warning. AI is a technological science that investigates and creates new theories, methods, technologies, and application systems to simulate and expand human intelligence. It is unavoidable to promote comprehensive cross-border integration of AI and aquaculture. The following will examine a variety of AI technology applications in aquaculture and investigate the feasibility of incorporating these methods into the design of an intelligent fish farm.

3.3.1. Water Quality Soft Measurement and Control Method

A Water quality in aquaculture has a significant impact on fish growth rate, health, and feed intake efficiency. A single water quality parameter in aquaculture water is easily influenced by other water quality parameters, increasing the difficulty of detection with a single detection method while also allowing for the use of soft measurement. The basic idea behind soft measurement technology is to infer or estimate important parameters that are difficult to observe using easily observed variables. The key components of soft measurement technology are feature extraction and the soft measurement model. Process mechanism models, statistically based machine learning models, and gray box models are the three main types of soft sensor modeling methods used today. Traditional water quality prediction methods have a slow convergence rate and low prediction accuracy. They are not appropriate for modeling with high dimensionality, small sample sizes, or parameter optimization influenced by subjective factors. Liu et al., (2013a, b) proposed two nonlinear aquaculture DO prediction methods using ACO-LSSVR and IPSO-LSSVR. The ACO-LSSVR method improves the ant colony optimization algorithm by employing local fine search based on the concept of "detection" and dynamic pheromone update, automates the acquisition of the best parameters of the LSSVR model, and constructs a DO nonlinear prediction model based on ACO and LSSVR. Compared to BPNN, this method reduces RMSE and running time by 67.9% and 2.3464 s, respectively. The IPSO-LSSVR method enhances the particle swarm optimization algorithm (IPSO) by employing an adaptive dynamic update strategy of inertial weights, enables fine search in the optimization process of LSSVR model combination parameters, and creates the DO nonlinear prediction model of IPSO fusion LSSVR. This method's RMSE and MAE are 29.36% and 67.46% lower than those of the traditional LSSVR. Water quality parameters in cage aquaculture are typically nonlinear, dynamic, changeable, and complex. Hu et al., (2019) developed a water quality prediction model based on a long short-term memory (LSTM) deep learning network, which repaired, corrected, and removed noise from water quality data using linear interpolation, smoothing, and moving average filtering techniques, and used Pearson's correlation coefficient to obtain correlation priors between pH, water temperature, and other water quality parameters. Short-term prediction accuracy of pH value and water temperature can reach 98.56% and 98.97%, with prediction times of 0.273 s and 0.257 s, respectively. The long-term prediction accuracy for pH value and water temperature was 95.76% and 96.88%, respectively.

There are numerous interrelated and nonlinear factors that influence the ammonia nitrogen content of pond culture. It is difficult to accurately predict ammonia nitrogen content using mathematical methods and traditional neural networks due to issues such as local convergence and low computational efficiency during data training. Chen et al., (2019) used the PCA method to screen the main factors influencing the change in ammonia nitrogen content, the wavelet threshold method to eliminate noise, and proposed an ammonia nitrogen prediction model based on the particle swarm optimization (PSO) algorithm and multivariable deep belief network (MDBN). This method's MAPE is 28.59% and 21.46% lower than that of traditional LSSVM and BPNN. Wang et al., investigated the production and nitrification of ammonia nitrogen in marine aquaculture water, using water temperature, salinity, pH value, and DO as auxiliary variables, and developed a soft measurement model of ammonia nitrogen concentration based on stochastic configuration networks (SCN) with fast convergence and strong generalization ability. The SCN method has an RMSE of less than 0.064 mg/L and a prediction time of less than 1.32 seconds. Yu et al., (2020) proposed a soft ammonia nitrogen measurement method using an empirical mode particle swarm optimization extreme learning machine (EMD-IPSO-ELM). This method decomposes the ammonia nitrogen inherently using the empirical mode method to obtain the number of hidden layer nodes of the extreme learning machine, and then selects the input layer weights and biases of the extreme learning machine using the particle swarm algorithm. On-line measurements of water quality and meteorological parameters allow for the estimation of ammonia nitrogen content. Rahim et al., (2020) used time series of physicochemical water quality variables such as pH, ORP, water temperature, and electrical conductivity in Greece's Small Prespa Lake to develop deep learning (CNN, LSTM, CNN-LSTM) and traditional machine learning models (support vector regression-SVR, decision tree-DT) for predicting DO and chlorophyll-a concentrations. The hybrid CNN-LSTM model combines the strengths of both standalone deep learning models, lowering the RMAE for DO prediction by 0.24% and 1.45% (for chlorophyll-a prediction by 6.41% and 6.48%) during the testing period when compared to CNN and LSTM models.

Ocean satellite remote sensing has a large observation range, a short repetition period, and high temporal and spatial resolution. It can image the entire world ocean in a short period of time, observe sea areas that ships cannot reach, and observe parameters that are difficult to measure using conventional methods. Large-scale farms can use existing and forecasted satellite remote-sensing data sets to assess environmental information such as chlorophyll-a, sea surface temperature, surface organic pollution, and ocean current (George, 2014) (Figure 3.3).

The success of designing affordable automated control systems for aquaculture will be widely applicable because it will improve water management. To improve distributed measurement, intelligent control, and centralized management of DO in a large-scale aquaculture pond, Ma et al., (2015) developed a pond DO cascade control system based on a variable universe fuzzy PID controller and an aerator speed PID controller.

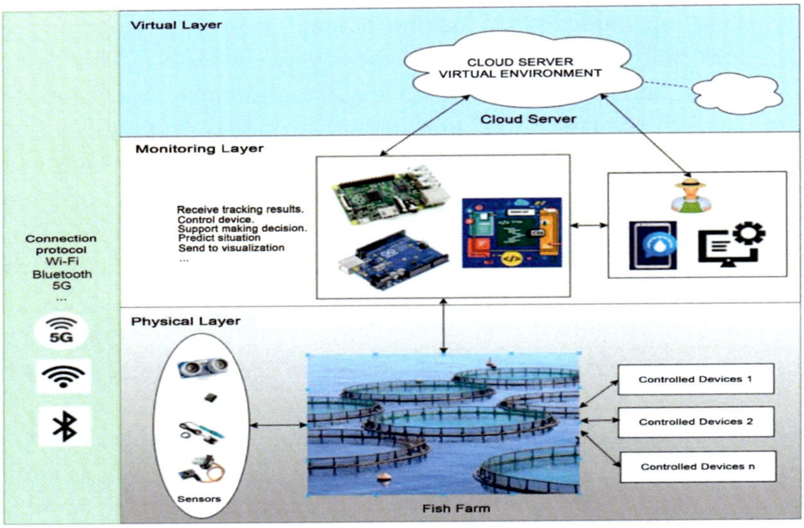

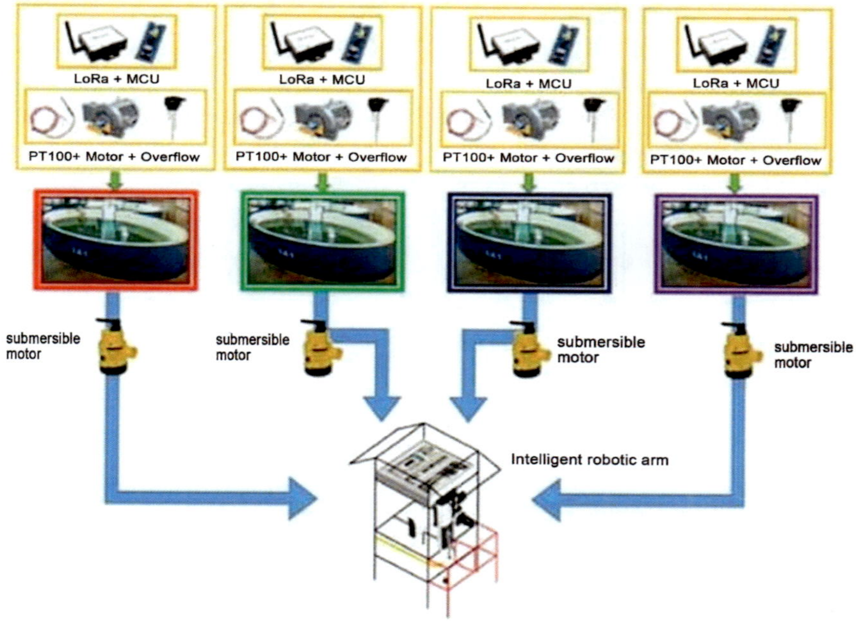

Figure 3.3. *Aquaculture water quality monitoring system.*

A variable universe fuzzy control unit tuned the PID controller parameters online in order to achieve the goal of DO adaptive control. Based on the change trend of the DO measurement sequences, a combined grey dissolved oxygen forecasting model is built using grey theory and weights. The predicted value serves as the feedback value for the variable universe fuzzy PID controller, allowing for predictive control and advanced DO regulation. Another predictive controller was developed for hourly control of DO to improve whiteleg shrimp production while minimizing costs; Marcos et al., (2017) first established a DO dynamic model by analyzing the process of DO production and consumption in shrimp ponds, as well as shrimp growth rate. The dynamic model is then run through the Nonlinear Extended Prediction Self-Adaptive Control (NEPSAC) algorithm to optimize the use of the aeration system. The NEPSAC algorithm is simple to integrate into RaspberryPi or Beaglebone to achieve edge computing control. Peng et al., (2018) used the differential evolution (DE) algorithm and a fuzzy controller to create a parallel joint control FLC model for the DO, ammonia nitrogen, and pH values of aquaculture water. The DE algorithm was used to modify the fuzzy rules of FLC. The FLC-DE model was used to verify water quality control at a shrimp breeding pond in Zhanjiang, China. The results indicate that the control accuracy for DO and pH was within ±0.25 mg/L and ±0.23, respectively. Model predictive control (MPC) is a kind of computational intelligence algorithm based on predictive model, which continuously scrolls and optimizes in a limited time domain, and achieves the optimal control of target by real-time feedback (Vrecko et al., 2011). Since the Dynamic Matrix Control (DMC) model predictive control algorithm is based on the nonlinear second-order rolling optimization, the amount of online calculation is not large, and it is more conducive to embedding into the local controller.

3.3.2. Intelligent Feeding Strategy

Bait accounts for up to 80% of total aquaculture costs. How to reduce feed costs is the key to maximizing aquaculture profits. Currently, the bait delivery method is primarily based on the fish's growth conditions and breeding experience, which makes accurate bait feeding impossible, and it is very easy to cause insufficient fish feeding, bait waste, and water pollution. The existing bait-casting machines provide insufficient information about the fish's feeding situation. Feeding back the evaluation results of the bait-feeding effect to the control system will allow you to adjust the feeding amount in real time. The artificial experience model is typically based on extensive aquaculture observation experience, and a regression fitting analysis method is used to establish a mathematical equation relating fish growth nutrient requirements to feed amount. According to Sun et al., (2016), fish feed requirements are based on their body mass/length ratio. The breeding environment's complexity and variability cause multiple disturbances in the artificial experience prediction model. Chen et al., (2020) proposed a feed intake prediction model for group fish in intensive aquaculture, which uses the MEA-BP neural network to automatically establish the mapping relationship between fish intake and environmental factors. In their study, the four variables of water temperature, DO, average fish weight, and fish number were used as inputs to BPNN, and GA (Genetic Algorithm) and MEA (mind evolution algorithm) algorithms were used to optimize the initial weights and thresholds of BPNN to improve matching accuracy. The linear correlation coefficient between the predicted value of the trained model and the measured value is 96%, with RMSE and MAPE of 6.89 and 0.04, respectively. Single morphological characteristics of fish bodies are insufficient for estimating fish feed demand. Some researchers use the biological energy of fish bodies, combined with the breeding environment, to predict fish feed demand. Digestional energy determines fish's dietary requirements at various stages of development. Liu et al., (2018) calculated the digestible energy of Carassius auratus gibelio by adding the expected energy gain, basic metabolic energy, heat increment of feeding, and metabolic energy of urine and gill, and then built a bioenergy model to predict feed intake. This model's predicted FCR value (1.47) remained consistent with the observed value (1.51) throughout the production cycle.

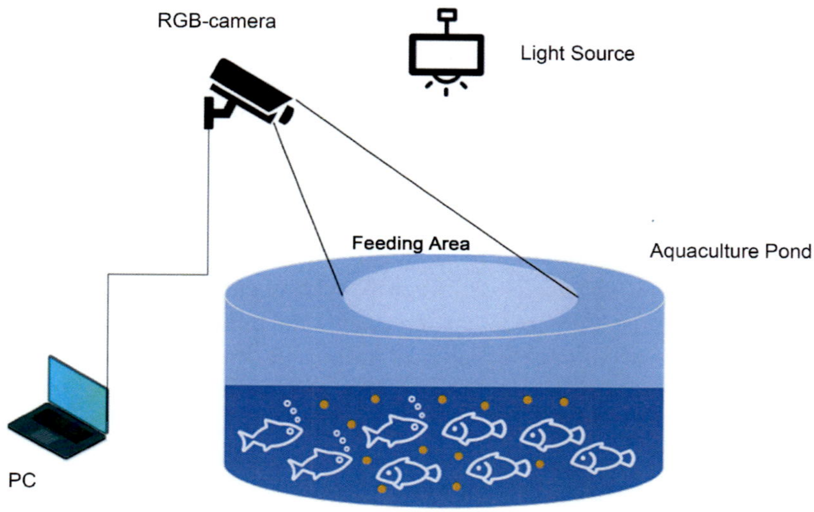

Figure 3.4. *Fish feeding behavior analysis.*

In recent years, advanced technologies such as machine vision, acoustic technology, and sensor data fusion have gradually applied to large-scale and refined aquaculture (Li et al., 2020a,b; Wu et al., 2015). Wang et al., (2015) proposed a method for accurately controlling and judging the reasonable feeding time, as well as improving feeding efficiency, by using machine vision technology to obtain the feeding parameters of the ingested fish. This method first detects the fish point set in the target area of the image sequence using the inter-frame difference method, then optimizes the fish point set to improve the efficiency of the least squares ellipse fitting algorithm, obtains the target fish area parameter value, and finally uses the parameter value change curve to calculate the maximum value of the area parameter. When the parameter value falls below the set threshold, the bait-casting machine is automatically turned off, allowing for intelligent control of fishery bait. Liu et al., (2014) used a machine vision system to study the feeding behavior of Atlantic salmon. To eliminate the effects of superimposed reflections on the water body's surface, the system used the inter-frame difference method, and the feeding intensity was assessed using the feeding activity index (CVFAI). The CVFAI-MOFAI mapping relationship model serves as the foundation for starting and stopping the automatic feeding machine. Researchers (Hu et al., 2015; Zhou et al., 2019) used machine vision technology to collect the aggregation degree of fish and water spray as the feature area of the image to describe the characteristic parameters of the feeding regularity. They proposed a variety of fish intake calculation models based on the change curve of the feeding regularity over time. However, as the fish gathering time and number of overlaps increase, the feeding model's accuracy may suffer. Zhou et al., (2018) used near-infrared machine vision to track fish movement patterns after feeding. Zhou et al., (2017a,b) developed an adaptive network-based fuzzy inference system (ANFIS) that uses feeding behavior parameters (FIFFB and SIFFB) as input variables to automate on-demand feeding. The ANFIS model can achieve 98% feeding decision accuracy. Furthermore, some researchers use the number of uneaten fish food pellets as a control inde□for starting and stopping the feeding machine in order to implement a feeding control strategy with fish as the main body. Foster et al., (1995) used a light-compensating camera (a Panasonic WV-B400 black and white CCD) to detect and count feed pellets in a sea cage during a feeding session. An automatic pellet counting algorithm based on recorded image sequences was created and tested. The average counting accuracy error is approximately 10%. Using the same procedure in a large glass-made water tank. Liu et al., (2015) created an adaptive method for detecting and identifying uneaten fish food pellets using Otsu Thresholding and a linear-time component labeling algorithm. The proposed algorithm for calculating the number of fish food pellets is more than 92% accurate, and it works well under non-uniform lighting conditions. Machine vision is vulnerable to the impact of the target environment during image collection, such as water surface reflections, fluctuations, and vibration noise. In order to improve monitoring accuracy and make the image signal more accurately reflect real-time feeding behavior, sensors with error correction functions must be included in feeding equipment designs.

Peixoto et al., (2020) found that sonar imaging technology is effective for monitoring marine fish. The fish's swimming position reflects how hungry it is. The underwater acoustic sensor monitors the fish's 3D movement trajectory and determines whether or not to stop feeding based on the detected fish position density. Zhou et al., (2017a,b) have demonstrated and applied this theory in practice. Sonar imaging technology, unlike underwater machine vision technology, does not need to consider the impact of light when detecting fish foraging behavior. However, the use of underwater sonar is sensitive to environmental noise. In the production process of a fish farm, environmental noise at the feeding area is usually distributed periodically, whereas feeding signals are random. As a result, using digital signal processing to remove environmental noise can help improve the method's decision-making accuracy.

The decision-making system for feeding fish in an intelligent fish farm should take into account the relationship between fish physiological and behavioral characteristics, the water quality conditions of the breeding environment, the feed formula, the operation method of the feeding equipment, and the growth rate of the fish. It is a more accurate way to create a fish feeding decision-making system based on dynamic balance models. To establish the mapping relationship between feed intake and the quantity of feed on offer, the system should use the multi-source data fusion method. When combined with high-efficiency intelligent control algorithms, the system can achieve adaptive feedback control for the feeding system's real-time decisions. Figure 3.5. illustrates the construction of an intelligent feeding system. Different information fusion configurations can be chosen for various intelligent fish farm scenarios.

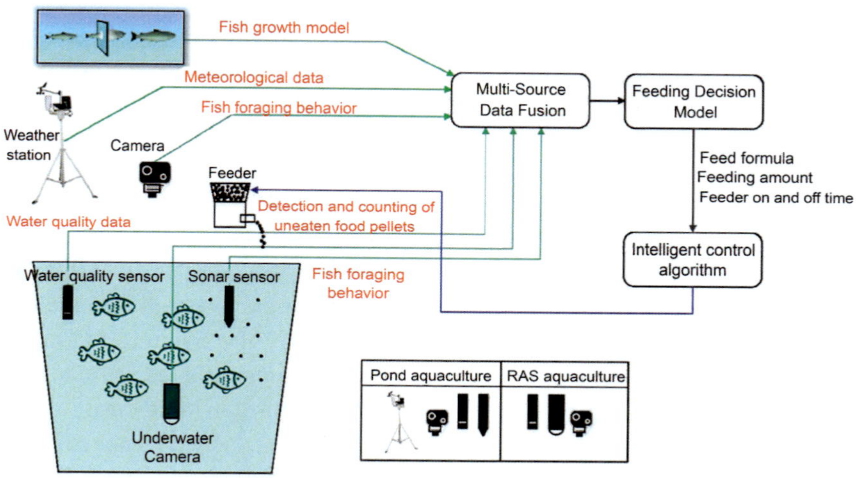

Figure 3.5. *Schematic view of intelligent feeding system.*

3.3.3. Behavior Analysis of Raised Species

The sensory, nervous, immune, and metabolic systems all have an impact on fish's stress response. Many factors influence it, including fish species, poison type, and exposure environment. Research on fish behavior can provide valuable theoretical data for aquaculture (Niu et al., 2018; Saberioon et al., 2016). Fish behavior in aquaculture is primarily concerned with swimming, feeding, reproductive, aggression, and carnivorous behavior.

Fish bioassay is one of the earliest biological monitoring methods, particularly for studying the effect of pollution fluctuations on fish behavior. Until now, indicators used to monitor and evaluate water quality have primarily included movement behavior (speed, height, turning times, and wagging frequency), respiratory behavior (gill cover movement frequency, respiratory rate, breathing depth, and cough frequency), and group behavior (dispersion, communication, average distance, and regional distribution). Kang et al., (2009) used Bio Fish Meter Lab equipment (Dual camera stereo cross distribution) to record three-dimensional (3D) data on Oryzias latipes movement, successfully resolving the signal plane crossing problem in video tracking technology. The abnormal behaviors, such as rapid swimming and surfacing, were used to determine whether the water quality was safe or not. Chew et al., (2009) developed a fully automated system to monitor the behavior of fish in real time, using machine vision collection and image analysis technology to investigate the activity levels of fish, the density of trajectory curves, avoidance behaviors, and social interactions within fish before and after water pollution, combined with individual abnormal monitoring indicators such as balanced loss rate, irregular turns, etc., to achieve comprehensive evaluation. The swimming behavior of fish before and after stress stimulation can be analyzed using a combination of heuristic algorithms and Markov models. Liu et al., (2011) used the self-organizing map (SOM) algorithm to classify si□types of zebrafish movement states before and after formaldehyde addition, and then used the hidden Markov model (HMM) to evaluate the Markov process of the various motion states. The early warning signal established using this method has an accuracy rate of 70.1-81.2. percent for estimating stress response behavior. In general image processing systems, tracking the movement patterns of a shoal of small fish is difficult. The fractal dimension is used to quantify the behavioral states of moving targets in a compressed manner, producing results that are consistent with individual behavior analysis. Multi-Layer Perceptron (MLP) training uses characteristic data describing various motion trajectories, such as speed, Y-axis coordinates, stop time, stop number, turning rate, and meandering. Ji et al., (2006) found that the MLP method can identify local information in movement patterns and accurately detect stress responses. Some non-contact bioelectric field technology can detect fish's ventilation frequency, cough rate, and heart rate. Acipenser baeri and Scophthalmus maximus had lower heart rates and higher respiratory amplitude and frequency when exposed to rapid hypoxia.

The study of fish feeding behavior can provide guidance information for precise feeding, water quality management, and the development and domestication of artificial compound bait, thereby improving fish feed utilization, lowering fish feed costs, and lowering water pollution levels. Using computer vision technology to monitor fish behavior and determine hunger levels can improve image processing ability and accuracy while also providing a theoretical basis for intelligent feeding. The corresponding cases have been mentioned several times in the preceding text. Understanding the law of fish reproductive behavior and studying the behavior development of larval fish can help guide artificial reproduction more accurately and improve the survival rate and quality of the process. Currently, artificial methods are commonly used to determine the se□ of specific fish. This method is labor-intensive, and the evaluation results are subjective. Hosseini et al., (2019) use

Deep Convolutional Neural Networks (DCNNs) and Support Vector Machines (SVMs) to automatically classify zebrafish based on their appearance, te□ture, and caudal fin color. The DCNNs algorithm accurately predicts the se□ ratio of the control group ($\varphi=0.97$), making it ideal for temperature induction e□periments ($\varphi=1.0$). However, the SVM method yields slightly lower results. The findings indicate that high ambient temperatures can reduce the color intensity of male zebrafish, and that the color of the male caudal fin is positively correlated with body weight and body length. Many fish make sounds for a variety of reasons, including courtship, mating, navigation, and environmental stress. Some groupers make species-specific courtship sounds while spawning and gathering. Using an HD video camera and audio separation software (Adobe Audition), Michelle et al., (2014) found that the black grouper's courtship sound has a peak frequency range of 67–96 Hz. Passive acoustic technology is more accurate than camera-based underwater audio monitoring and can be used to observe fish reproductive cycles. Underwater acoustic research will focus on combining discrete wavelet transform (DWT) with deep learning techniques like CNN, RNN, and LSTM for signal denoising, feature extraction, and automatic classification (Ibrahim et al., 2018).

Fish, like other animals, often defend their food and territory. Aggressive behavior can injure fish's skin and fins, making them susceptible to disease and death. Aggressive behavior also consumes energy for growth, resulting in reduced aquaculture production, lower food conversion efficiency, and slower growth. Gutiérrez et al., (2018) used computer vision technology to create a novel device for automatic monitoring and quantification of zebrafish aggression. ZebraLab software measured the "ratio" and "angle" of aggressive behavior in real-time during drug treatment and mirror-induced aggression e□periments. The intensity of aggressive behavior was quantified using the H inde□ ($H=angle*\log(ratio)/2$). Photoperiod, light intensity, individual differences, different growth stages, food adequacy and palatability, rearing density, temperature, and bottom quality all influence the intensity of fish cannibalism. Domesticated bait, adequate and evenly distributed food pellets, appropriate rearing density, and timely grading can help to reduce mutual cannibalization among carnivorous fish.

Deep learning is the most advanced machine learning method available today. It stems from research into artificial neural network architecture, which has a large number of hidden layers and millions of parameters. Deep learning combines low-level features to create more abstract high-level representation attribute categories or features in order to determine the distributed feature representation of data. Deep learning has produced remarkable results in a variety of fields, including image understanding, video, and audio analysis. Intelligent fish farms use machine vision (3D observation structure), sonar detection, and deep learning technology to analyze the behavior of cultured animals in real time. Figure 3.6. illustrates the design of a fish behavior monitoring platform.

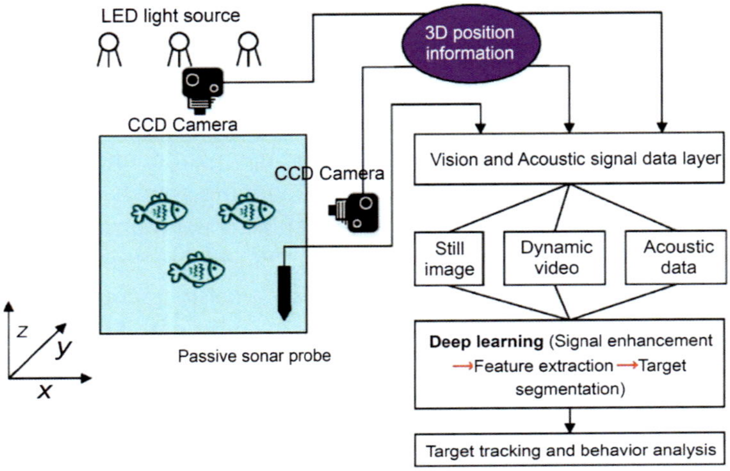

Figure 3.6. *Design of fish behavior monitoring and analysis platform.*

3.3.4. Biomass Statistics

The biomass statistics are critical in supporting fish farmers' decisions such as fish food dosage, drug consumption, and fish mortality. It is typically used to collect subsamples from the culture pond, weigh them, and then calculate the biomass of the entire pond. This method is highly error-prone and labor-intensive, resulting in a high stress response and fish mortality. Machine vision has made significant contributions to online non-destructive biomass estimation in aquaculture (Li et al., 2019; Dios et al., 2003; Li et al., 2020a, b). Fish weight is closely related to their length, width, area, and circumference at different growth stages (Shi et al., 2020). These parameters will serve as a key foundation for estimating fish biomass. Viazzi et al., (2015) used a 2D computer vision technique to estimate the mass of Jade perch (Scortum barcoo) after photographing fish samples of various sizes out of water. The image processing algorithm was used to extract the fish's length, area, and contour information from the sequence photographs. Regression analysis was used to create a linear model with input variables area, length, and width and output variable weight. The weight of Jade perch can be accurately determined using a single factor regression equation that excludes the fin tail. The model's correlation coefficient (R2) is 0.99. The binocular stereo vision system generally obtains two digital images of the measured object simultaneously from different angles using dual cameras, and restores the object's 3D geometric information using the parallax principle, reconstructing the object's 3D contour and position. A deformable adaptive computer vision algorithm can automatically fit a Bluefin Tuna's ventral silhouette while swimming. Pau et al., (2018) used local thresholding technology to perform image segmentation and extract a single Bluefin Tuna object from the stereo vision system's video frame. The deformable tuna model is defined as a vector of six parameters that includes the positions of 18 vertebral points and 35 profile points from the fish's mouth to the end of the caudal peduncle keel. Euclidean distance, snout fork length, and five widths in various sections of the fish silhouette were used to investigate fattening evolution. Although machine vision methods have the advantage of remote non-interference operation in monitoring fish biomass, their detection accuracy suffers from poor visibility, lighting changes, distance and relative direction changes between camera and fish, swimming speed, and rearing density in aquaculture. Fast R-CNN and Faster R-CNN have excelled in object detection research (Yang et al., 2020). However, these algorithms require a complex execution pipeline to achieve high recognition accuracy and low FPS. To improve results for detecting different fish, Raza et al., (2020) increased the YOLOv3 detection scale to 4, used K-means clustering to increase the anchor box, developed a novel transfer learning technique, and improved the loss function. The improved YOLOv3 model has an average accuracy of 91.30%, which outperforms Faster R-CNN, YOLOv2, and the original YOLOv3. The method's detection speed is approximately 1000 times faster than R-CNN and 100 times faster than Fast R-CNN. The improved YOLOv3 model can also detect objects in real time.

Another noninvasive monitoring method for estimating fish biomass in real time is laser scanning. Fish biomass is typically calculated by multiplying the density (ρ) and volume (V) of the fish. Almansa et al., (2015) investigated a laser scanning method for estimating the total biomass of Senegalese sole by measuring the volume of all Senegalese sole in the culture pond. The laser scanning system's digital camera and structured light can move at the same time, with the distance and inclination angle remaining constant. As modeling data, the digital camera records the laser line reflected from the surface of the fish body. The fish layer thickness at each point was determined by comparing the height difference of the laser beam in spatial coordinates. Under medium and high stocking density conditions with large fish, the ratio of fish biomass to fish layer volume (FB/FLV) was very close to one. This method requires a very low interstitial water content when the monitoring objects are overlapped, making it ideal for benthic and relatively inactive species such as the turbot and tongue sole families. When the laser scanner performs automatic image analysis, uneven illumination and unnecessary noise can cause image distortion.

However, this can be handled for on-site biomass estimation by allowing regular and frequent repeated operations to discard bad images and improve scanning accuracy.

In comparison to the limited penetration ability of light in water, the attenuation of sound waves is much lower. Identification sonar is a type of multi-beam system that uses an acoustic lens to transmit an independent beam. It can produce high-definition images that are nearly identical to optical images in dark or turbid water areas, allowing it to recognize targets. Installing an echosounder with two floating split-beam transducers allows you to estimate the biomass in Gilthead sea bream and sea bass cages with varying rearing densities and individual sizes. Using this platform, Djemali et al., (2017) assessed fish biomass through fish counting rather than echo integration. Methane gas bubbles and fish behavior in deep sea cages may influence fish abundance estimates. Seasons and photoperiods also serve as important reference points for developing acoustic sampling strategies. To track and count moving fish, use dual frequency recognition sonar (DIDSON) and a real-time image processing algorithm. Jing et al., (2017) used the depth first search algorithm to find a single fish target, which they then tracked using the e□tended Kalman filter algorithm and the nearest neighbor algorithm. The huge randomness of the dense target's motion accounts for 16% of the statistical error in fish numbers.

Modern advanced remote sensing technology, which uses satellite remote sensing images and relevant professional software to analyze marine fishery resources, can accurately determine the position of fish stocks, significantly improving fishing accuracy and quantity. Visible light remote sensing technology measures water color changes in fish farms, estimates chlorophyll-a concentrations in sea water, and evaluates the distribution of marine fish stocks fed on zooplankton and phytoplankton (George, 2014). Mahendra et al., (2010) used thermal infrared remote sensing to study fish aggregation and migration, as well as assess coral reef health. Sea surface height remote sensing technology is a microwave remote sensing method that uses a satellite altimeter to collect abnormal sea surface height data and then analyzes the fish farm environment. To detect fish stocks, altimeter data is used to determine the position of oceanic frontal regions where marine life tends to gather (Digby et al., 1999).

The evaluation of fish resources in an intelligent fish farm is an automated, nondestructive, and rapid analysis process. Machine vision, sonar, and satellite remote sensing technology all have potential applications. Video sensors allow for easy e□ternal morphometry of fish (Aguzzi et al., 2009). Deep learning techniques like R-CNN, YOLO, and SSD can help identify and count species in farmed areas (Yang et al., 2021). The platform's basic design concept is the same as shown in Figure 3.6. Furthermore, the system can integrate satellite remote sensing monitoring sensors to obtain large-scale, high-frequency, continuous, and objective data sets. Given the challenges of dense underwater targets and a complex environment, software implementation methods for machine vision must be optimized from image preprocessing to target recognition, multi-target tracking algorithms, and counting rules. The combination of satellite remote sensing technology and models aids in the management of fishery resources at the ecosystem level.

3.3.5. Fault Diagnosis of Equipment

Abnormal working conditions will reduce the availability and integrity of the aquaculture IoT measurement and control cluster system, resulting in fatal aquaculture accidents. When a fault occurs, it is difficult to determine the type, severity, and location of the fault without diagnostic information. Powering off and shutting down is frequently required for disassembly and inspection, which slows down the manufacturing process. The working condition data can be used to obtain mathematical model-based diagnosis information, allowing faults to be identified and located without turning off the power and disassembling the system for inspection. Yang et al., (2017) propose a fault feature extraction method based on DPCA-VMD-SVD, which addresses the complexity of multivariate combination feature extraction, strong autocorrelation of variables, and significant non stationarity of fault conditions. Based on dynamic principal component analysis (DPCA), the sample's basic matrix is mapped into the dynamic principal component space, and 31 dynamic principal components are chosen based on their proportion of cumulative variance. The dynamic principal component matrix is then decomposed into three intrinsic mode functions (IMF) using variational mode decomposition (VMD). The IMF matrix of each principal component is compressed into a vector of singular values using singular value decomposition (SVD). A 3×31 sample feature matrix is then extracted from a 30×21 sample basis matrix. To address the challenges of small sample size, diverse categories, inseparable linearity, and classification conflicts in aquaculture IoT, the researchers used cross-correlation of pattern classes to create 24 sets of SVM two-class classifiers. Multi-objective parameter optimization was performed using swarm intelligence optimization algorithms, and the classification conflict was resolved using the D-S evidence theory, which increased classification accuracy by 2.5. to 4.4 percentage points. The computer monitoring system must consider its reliability in order to prevent fish deaths caused by an excess of certain environmental factors due to a control system failure. The fault tree analysis (FTA) is an important method for assessing system reliability. The system's undesirable events are used as analysis targets to determine the logical relationship between potential component failures, software defects, human errors, and system failures, as represented by an inverted tree diagram. Chen et al., (2017) found that combining FTA and FNN methods can enable intelligent fault diagnosis in aquaculture IoT systems. By identifying the system's weak links, the system design was optimized to ensure normal operation in aquaculture. The fault tree depicted the causes of fault events and their logical relationships, while the FNN was used to train the relationship mapping between fault symptoms and faults. Liu et al., (2020) found that an expert system using machine vision and surveillance cameras to detect anomalies in aerators is cost-effective, easy to promote, and highly automated. Long-distance aerator targets can be detected from various positions by performing a three-step operation that includes maximum contour detection, candidate region detection, and object region detection. In the working state detection module, the Kanade-Lucas-Tomasi (RF-KLT) algorithm, which is based on region construction and corner matching, can extract robust motion features in a fixed area, broadening the application scope of traditional optical flow methods and breaking through the limitation of adjacent frames. When the system monitors all weather conditions, infrared cameras and improved machine vision processing algorithms appear to be essential.

Modern mechanical equipment in intelligent fish farms is becoming more automated, comple□, efficient, and intelligent. The cluster system for monitoring and diagnosis has the following characteristics: large scale, multiple measuring points, and high sampling frequency. The measurement data are the result of the coupling effects of multiple physical quantities over different time periods in various parts of the monitored object. Mechanical equipment has been in normal working order for a long time, with abnormal conditions accounting for only a small portion. Machine condition monitoring and fault diagnosis must be quick and accurate to avoid equipment damage or production disruptions caused by abnormal data loss. As a result, traditional fault diagnosis methods based on feature extraction and pattern recognition are no longer adequate for this application. Deep learning, a breakthrough in modern AI, can automatically learn valuable features from the original feature set or even raw data, reducing reliance on advanced signal processing technology, artificial feature e□traction, and cumbersome feature selection technology. The fault diagnosis methods developed using the deep learning framework will undoubtedly become the dominant trend in mechanical equipment fault diagnosis in intelligent fish farms.

Figure 3.7 shows an intelligent fault diagnosis framework for aquaculture IoT. There are two stages to the diagnosis process. To begin, all data must be preprocessed; feature parameters that can represent fault symptoms are extracted using deep learning; and a specific number of sample sets are chosen to train the neural network to produce the e□pected diagnosis network and classifier. Second, the trained neural network and classifier diagnose the system's online data.

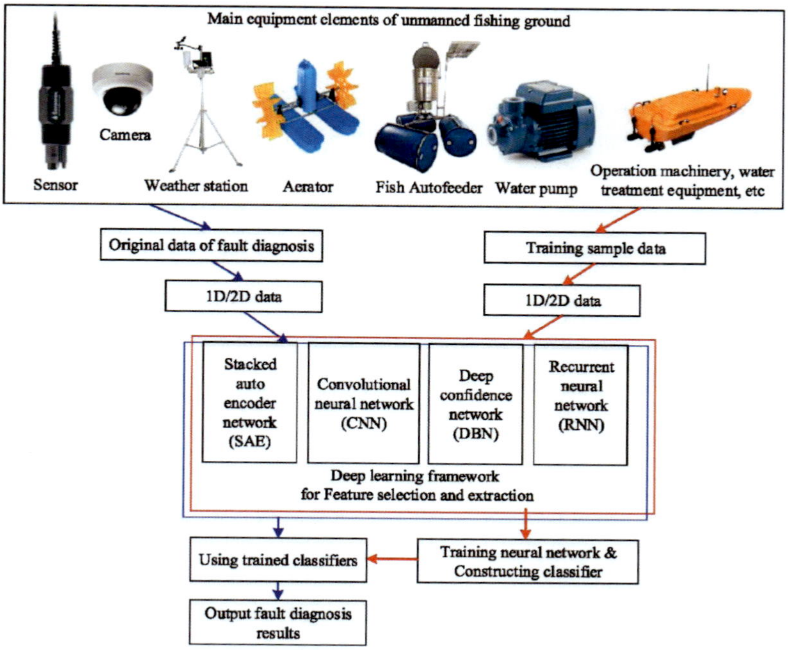

Figure 3.7. *Deep learning fault diagnosis based on multi-source heterogeneous data fusion.*

3.4. CHALLENGES

Based on the advanced level of information technology, the intelligent fish farm can be roughly divided into three stages: In the early stages, most of the work in a fish farm is dependent on aquaculture experts who can operate and control remotely. In the intermediate stage, the IoT system can operate autonomously, but still requires a small team to execute commands and make production decisions, known as the unattended fish farm. In the advanced stage, the fish farm's production does not require any human participation at all; all operations and management businesses are independently planned and decided by the cloud management platform, and robots and intelligent equipment operate autonomously, resulting in a completely unmanned farm. The intelligent fish farm can use intelligent digital technologies to address issues such as aquaculture labor shortages, water pollution, high risk, and low efficiency, but some technical details must be considered.

With the deep integration of agricultural technology and advanced information technology such as IoT, cloud computing, big data, and AI, agricultural robots, a new generation of intelligent agricultural machinery, will break through innovation bottlenecks and become widely used. The future fish farm will require a large number of agricultural robots. Deep learning, new materials, human-computer integration, and haptic feedback are all examples of new agricultural robot technologies under investigation. Because of the complex and changing working environment in intelligent fish farms, as well as the variety of cultured species and their varying growth states, aquaculture robots face significant challenges in the areas of target identification and positioning algorithm optimization, navigation and path planning algorithm optimization, and operation object sorting and monitoring algorithm. IoT technology is critical in the collaborative efforts of various production factors in an intelligent fish farm. To achieve sustainable development, aquaculture IoT must address three issues. Water quality sensors are susceptible to surface attachment of biological elements and have low online measurement reliability. The sensors have a short life span and low intelligence. The environmental complexity of intelligent fish farms, combined with the low power consumption of sensors, has resulted in increased requirements for agricultural IoT data transmission. In terms of network transmission security technology, anti-interference technology, and automatic dynamic networking technology with low power consumption, there is still a significant gap when compared to industrial process measurement and control. The instability of network transmission creates challenges for back-end data processing and intelligent data analysis. The standards for supporting agricultural IoT are lagging behind, and there is a lack of unified guidance and specification documents for the design of sensor network architecture in fish farms. Designers of IoT devices typically define private data transfer protocols. There is no uniform standard to follow when applying sensor data fusion and designing upper application systems, which is detrimental to the advancement of industrial technology.

In the context of aquaculture big data, the biodiversity of aquatic animals and the comple□ity of their growing environment in intelligent fish farms present challenges for data collection. Many studies are currently being conducted in laboratory settings. Traditional video image acquisition, which is used to monitor the occurrence of fish diseases and abnormal fish behavior in natural environmental conditions, has low accuracy, which has always been a bottleneck for accurate aquaculture big data collection. The combination of big data technology and aquaculture only applies existing intelligent methods in a simple and direct manner, with little regard for aquaculture characteristics, resulting in the practicability of aquaculture big data analysis falling behind market demand. To increase the level of intelligence in the aquaculture industry, key intelligence technologies must be advanced, as well as the depth and breadth of applications of deep learning, knowledge computing, swarm intelligence, and hybrid-augmented intelligence in aquaculture. Furthermore, current aquaculture big data research focuses solely on a single or partial production process, lacks horizontal and vertical relevance, and provides no solution for the entire aquaculture industry.

Deep learning is the key breakthrough in AI for intelligent fish farms. Deep learning, as opposed to traditional machine learning, can extract more features from agricultural images and structured data, as well as effectively combine with agricultural machinery to aid in the development of intelligent aquaculture equipment. There are some deficiencies in the application of deep learning in aquaculture, as follows: To begin, deep learning requires large data sets for model training, verification, and testing, necessitating the development of cameras or sensor equipment to collect data in a variety of environments. The ambiguity and instability of underwater imaging systems, as well as the high frequency disturbance and drift of sensors, add to the design complexity of deep learning algorithms. Second, the majority of deep learning-based aquaculture problems require supervised learning, which necessitates labeling of the sample data. In general, more professionals are required to participate and manually mark the target category. Finally, while deep learning can learn the features of training data sets well, it cannot generalize beyond the expression ability of the data set, implying that the migration of deep learning algorithms requires additional consideration.

3.5. CONCLUSION

The aquaculture revolution must shift from traditional experience aquaculture to digital intelligent aquaculture due to issues such as degradation of offshore fish stocks, marine pollution (Sarma et al., 2013), older agricultural workers, and a decrease in aquaculture employees. Modern emerging technologies such as AI, big data, IoT, sensors, machine vision, and robots will gradually integrate into the entire aquaculture production process in order to completely liberate traditional labor force, and eventually realize multi-scene all-weather real-time monitoring of production environment, big data analysis based on cloud platform, and real-time intelligent decision-making.

This chapter introduces a new concept of intelligent fish farms and briefly reviews the use of new generation information technology in aquaculture. The application of intelligent measurement and control, feeding, inspection, and harvesting equipment in intelligent fish farms is then detailed. Based on the analysis and summary of the intelligent fish farm's core application layer, some subsystem integration schemes are proposed. The proposal and construction of an intelligent fish farm has a certain reference value and guiding significance for future digital, intelligent, and unmanned aquaculture development. The construction of an intelligent fish farm is far more complicated than other intelligent farming projects. The reliability and service life of sensors, the robustness and accuracy of analysis and decision-making models, the reliability of data transmission via IoT technology, and the cooperation efficiency of various aquaculture intelligent equipment must all be addressed.

In the future, the rapid development of intelligent fish farms is ine□tricably linked to sound infrastructure construction and strong government policy support. The government should encourage potential aquaculture enterprises to upgrade first, introducing advanced agricultural technology and high-tech talents, while also ensuring the healthy and environmentally friendly operation of intelligent fish farms by regulating aquaculture sustainability indicators.

REFERENCES

1. Aguzzi, J., Costa, C., Fujiwara, Y., Iwase, R., Ramirez-Llorda, E., & Menesatti, P. (2009). A novel morphometry-based protocol of automated video-image analysis for species recognition and activity rhythms monitoring in deep-sea fauna. *Sensors, 9*, 8438–8455. https://doi.org/10.3390/s91108438.
2. Al-Hussaini, K., Zainol, S. M., Ahmed, R. B., & Daud, S. (2018). IoT monitoring and automation data acquisition for recirculating aquaculture system using fog computing. *Journal of Computational Hardware Engineering, 1*, 1–13. https://doi.org/10.63019/jche.v1i2.610.
3. Almansa, C., Reig, L., & Oca, J. (2015). The laser scanner is a reliable method to estimate the biomass of a Senegalese sole (*Solea senegalensis*) population in a tank. *Aquaculture Engineering, 69*, 78–83. https://doi.org/10.1016/j.aquaeng.2015.10.003.
4. Barbedo, J. G. A. (2014). Computer-aided disease diagnosis in aquaculture: Current state and perspectives for the future. *Revista Innover, 1*, 19–32.
5. Bonadies, S., Smith, N., Niewoehner, N., Lee, A. S., Lefcourt, A. M., & Gadsden, S. A. (2018). Development of proportional-integral-derivative and fuzzy control strategies for navigation in agricultural environments. *Journal of Dynamic Systems, Measurement, and Control, 140*, 61007–610012. https://doi.org/10.1115/1.4038504.
6. Brugère, C., Aguilar-Manjarrez, J., Beveridge, M. C. M., & Soto, D. (2019). The ecosystem approach to aquaculture 10 years on—A critical review and consideration of its future role in blue growth. *Reviews in Aquaculture, 11*, 493–514. https://doi.org/10.1111/raq.12242.
7. Chen, L., Yang, X. T., Sun, C. H., Wang, Y. Z., & Xu, D. M., & Zhou, C. (2020). Feed intake prediction model for group fish using the MEA-BP neural network in intensive aquaculture. *Information Processing in Agriculture, 7*, 261–271. https://doi.org/10.1016/j.inpa.2019.09.001.
8. Chen, Y. Y., Cheng, Y. J., Yang, L., Liu, Y. Q., & Li, D. L. (2019). Prediction model of ammonia-nitrogen in pond aquaculture water based on improved multi-variable deep belief network. *Transactions of the Chinese Society of Agricultural Engineering, 35*, 195–202. https://doi.org/10.11975/j.issn.1002-6819.2019.07.024.
9. Chen, Y. Y., Zhen, Z. M., Yu, H. H., & Xu, J. (2017). Application of fault tree analysis and fuzzy neural networks to fault diagnosis in the Internet of Things (IoT) for aquaculture. *Sensors, 17*, 153–167. https://doi.org/10.3390/s17010153.

10. Chew, B. F., Eng, H. L., & Thida, M. (2009). Vision-based real-time monitoring on the behavior of fish school. In *Proceedings of the IAPR Conference on Machine Vision Applications (MVA 2009)* (pp. 90–93). Yokohama, Japan.
11. Digby, S., Antczak, T., Leben, R., Born, G., Barth, S., Cheney, R., Foley, D., Goni, G. J., Jacobs, G., & Shay, N. (1999). Altimeter data for operational use in the marine environment. *Oceans '99 MTS/IEEE Conference Proceedings: Riding the Crest into the 21st Century, 2*, 605–613. https://doi.org/10.1109/OCEANS.1999.804771
12. Dios, J. R. M., Serna, C., & Ollero, A. (2003). Computer vision and robotics techniques in fish farms. *Robotica, 21*, 233–243. https://doi.org/10.1017/S0263574702004733.
13. Djemali, J., & Laouar, H. (2017). Acoustic fish biomass assessment in a deep Tunisian reservoir: Effects of season and diel rhythm on survey results. *African Journal of Aquatic Science, 42*, 35–43. https://doi.org/10.2989/16085914.2016.1277181.
14. Food and Agriculture Organization of the United Nations. (2020). *The State of World Fisheries and Aquaculture 2020: Sustainability in Action.* Rome: FAO.
15. Foster, M., Petrell, R., Ito, M. R., & Ward, R. (1995). Detection and counting of uneaten food pellets in a sea cage using image analysis. *Aquaculture Engineering, 14*, 251–269. https://doi.org/10.1016/0144-8609(94)00006-M.
16. Francesca, A., & Corrado, C. (2020). Precision aquaculture: A short review on engineering innovations. *Aquaculture International, 28*, 41–57. https://doi.org/10.1007/s10499-019-00443-w.
17. George, G. (2014). Numerical modelling and satellite remote sensing as tools for research and management of marine fishery resources. In *Remote Sensing and Modelling: Advances in Coastal and Marine Resources* (pp. 431–452). Coastal Research Library. https://doi.org/10.1007/978-3-319-06326-3_18.
18. Gopalakrishnan, A., George, G., & Ignatius, B. (2017). Supporting the livelihood of four million fisher folk in Indian marine fisheries—ICAR-CMFRI's recent initiatives. In *11th Indian Fisheries and Aquaculture Forum* (23 November 2017). Kochi.
19. Gutiérrez, H. C., Vacca, I., Pons, A. I., & Norton, W. K. J. (2018). Automatic quantification of juvenile zebrafish aggression. *Journal of Neuroscience Methods, 296*, 23–31. https://doi.org/10.1016/j.jneumeth.2017.12.012.
20. Hosseini, S., Simianer, H., Tetens, J., Brenig, B., Herzog, S., & Sharifi,

A. R. (2019). Efficient phenotypic se□ classification of zebrafish using machine learning methods. *Ecology and Evolution, 9*, 13332–13343. https://doi.org/10.1002/ece3.5788.

21. Hu, J., Li, D. L., Duan, Q. L., Han, Y. Q., Chen, G. F., & Si, X. L. (2012). Fish species classification by color, texture and multi-class support vector machine using computer vision. *Computers and Electronics in Agriculture, 88*, 133–140. https://doi.org/10.1016/j.compag.2012.07.008.
22. Hu, L. Y., Wei, Y. Y., & Zheng, J. H. (2015). Research on intelligent feeding method based on machine vision technology. *Journal of Tropical Oceanography, 34*, 90–95. https://doi.org/10.3969/j.issn.1009-5470.2015.04.013.
23. Hu, Z. H., Zhang, Y. R., Zhao, Y. C., Xie, M. S., Zhong, J. Z., Tu, Z. G., & Liu, J. T. (2019). A water quality prediction method based on the deep LSTM network considering correlation in smart mariculture. *Sensors, 19*, 1420. https://doi.org/10.3390/s19061420.
24. Huan, J., Li, H., Wu, F., & Cao, W. J. (2020). Design of water quality monitoring system for aquaculture ponds based on NB-IoT. *Aquaculture Engineering, 90*, 102088–102097. https://doi.org/10.1016/j.aquaeng.2020.102088.
25. Ibrahim, A. K., Zhuang, H., Chérubin, L. M., Schärer-Umpierre, M. T., & Erdol, N. (2018). Automatic classification of grouper species by their sounds using deep neural networks. *Journal of the Acoustical Society of America, 144*, EL196–EL202. https://doi.org/10.1121/1.5054911.
26. Ji, C. W., Lee, S. H., Kwak, I. S., Cha, E. Y., & Chon, T. S. (2006). Computational analysis of movement behaviors of medaka (*Oryzias latipes*) after the treatments of copper by using fractal dimension and artificial neural networks. *Environmental Toxicology, 10*, 93–107. https://doi.org/10.2495/ETOX060101.
27. Jing, D. X., Han, J., Wang, X. D., Wang, G. Y., Tong, J. F., Shen, W., & Zhang, J. (2017). A method to estimate the abundance of fish based on dual-frequency identification sonar (DIDSON) imaging. *Fisheries Science, 83*, 685–697. https://doi.org/10.1007/s12562-017-1111-3.
28. Kang, I. J., Moroishi, J., Yamasuga, M., & Kim, S. G., & Oshima, Y. (2009). Swimming behavioral toxicity in Japanese medaka (*Oryzias latipes*) exposed to various chemicals for biological monitoring of water quality. In *Atmospheric and Biological Environmental Monitoring*. Springer, Dordrecht. https://doi.org/10.1007/978-1-4020-9674-7_20.
29. Lee, K. H., Lee, D., Yoon, J., Kwon, O., & Lee, J. (2017). Differential pH-sensitive sensor with dc electrode-offset compensation. *Electronics*

Letters, 53, 251–253. https://doi.org/10.1049/el.2016.3330Ci.

30. Li, D. L., Hao, Y. F., & Duan, Y. Q. (2019). Nonintrusive methods for biomass estimation in aquaculture with emphasis on fish: A review. *Reviews in Aquaculture, 12*, 1390–1411. https://doi.org/10.1111/raq.12388.
31. Li, D. L., Miao, Z., Peng, F., Wang, L., Hao, Y. F., Wang, Z. H., Chen, T., Li, H., & Zheng, Y. Y. (2020). Automatic counting methods in aquaculture: A review. *Journal of the World Aquaculture Society, 51*, 1–15. https://doi.org/10.1111/jwas.12745.
32. Li, D. L., Wang, Z. H., Wu, S. Y., Miao, Z., Du, L., & Duan, Y. Q. (2020). Automatic recognition methods of fish feeding behavior in aquaculture: A review. *Aquaculture, 528*, 735508–735518. https://doi.org/10.1016/j.aquaculture.2020.735508.
33. Liu, H. Y., Xu, L. H., & Li, D. W. (2015). Detection and recognition of uneaten fish food pellets in aquaculture using image processing. *6th International Conference on Graphic and Image Processing (ICGIP), 9443*. https://doi.org/10.1117/12.2179138.
34. Liu, S. Y., Tai, H. J., Ding, Q. S., Li, D. L., Xu, L. Q., & Wei, Y. G. (2013). A hybrid approach of support vector regression with genetic algorithm optimization for aquaculture water quality prediction. *Mathematical and Computer Modelling, 58*, 458–465. https://doi.org/10.1016/j.mcm.2011.11.021.
35. Liu, S. Y., Xu, L. Q., Li, D. L., Li, Q. C., Jiang, Y., Tai, H. J., & Zeng, L. H. (2013). Prediction of dissolved oxygen content in river crab culture based on least squares support vector regression optimized by improved particle swarm optimization. *Computers and Electronics in Agriculture, 95*, 82–91. https://doi.org/10.1016/j.compag.2013.03.009.
36. Liu, X. J., Guo, X., Wang, C. F., Li, D. P., & Dominique, B. (2018). Bioenergetics-based model to determine growth, feed requirement and waste output of giber carp (*Carassius auratus* gibelio). *Acta Hydrobiologica Sinica, 42*, 221–231. https://doi.org/10.7541/2018.028.
37. Liu, Y. Q., Yu, H. H., Gong, C. Y., & Chen, Y. Y. (2020). A real time expert system for anomaly detection of aerators based on computer vision and surveillance cameras. *Journal of Visual Communication and Image Representation, 68*, 102767–102782. https://doi.org/10.1016/j.jvcir.2020.102767.
38. Liu, Y., Lee, S. H., & Chon, T. S. (2011). Analysis of behavioral changes of zebrafish (Danio rerio) in response to formaldehyde using self-organizing map and a hidden Markov model. *Ecological Modelling, 222*, 2191–2201. https://doi.org/10.1016/j.ecolmodel.2011.02.010.

39. Liu, Z. Y., Li, X., Fan, L. Z., Lu, H. D., Liu, L., & Liu, Y. (2014). Measuring feeding activity of fish in RAS using computer vision. *Aquaculture Engineering, 60*, 20–27. https://doi.org/10.1016/j.aquaeng.2014.03.005.
40. Ma, C. G., Zhao, D. A., Wang, J. G., Chen, Y. J., & Li, Y. Z. (2015). Intelligent monitoring system for aquaculture dissolved oxygen in pond based on wireless sensor network. *Transactions of the Chinese Society of Agricultural Engineering, 31*, 193–200. https://doi.org/10.3969/j.issn.1002-6819.2015.07.028.
41. Mahendra, R. S., Bisoyi, H., Prakash, C. M., Velloth, S., Sinivasa Kumar, T., & Nayak, S. (2010). Applications of the multi-spectral satellite data from IRS-P6 LISS-III and IRS-P4 OCM to decipher submerged coral beds around Andaman Islands. *International Journal of Earth Sciences and Engineering, 3*, 626–631.
42. Malik, S., Kumar, T., & Sahoo, A. K. (2017). Image processing techniques for identification of fish disease. *IEEE 2nd International Conference on Signal and Image Processing (ICSIP)*, Singapore, 55–59. https://doi.org/10.1109/SIPROCESS.2017.8124505
43. Marcos, S., Ismael, S., & Edilberto, V. (2017). Adaptive predictive control of dissolved oxygen concentration in a dynamic model of white leg shrimp culture. *Chilean Conference on Electrical, Electronics Engineering, Information and Communication Technologies*. https://doi.org/10.1109/CHILECON.2017.8229499.
44. Martin, F., Kevin, F., Tomas, N., Eirik, S., Jo, A. A., Tim, D., Harkaitz, E., Win, W., Annette, S., Leif, M. S., Christian, S., Kristoffer, R. S., & Morten, O. A. (2018). Precision fish farming: A new framework to improve production in aquaculture. *Biosystems Engineering, 173*, 176–193. https://doi.org/10.1016/j.biosystemseng.2017.10.014.
45. Mateo-Sagasta, J., Marjani, S., Turral, H., & Burke, J. (2017). Water pollution from agriculture: A global review executive summary. Rome: FAO and IWMI.
46. Michelle, T. S., Michael, N., Timothy, J. R., & Richard, S. A. (2014). Sounds associated with the reproductive behavior of the black grouper (*Mycteroperca bonaci*). *Marine Biology, 161*, 141–147. https://doi.org/10.1007/s00227-013-2324-3.
47. Moges, F. D., Patel, P., Parashar, S., & Das, B. (2019). Mechanistic insights into diverse nano-based strategies for aquaculture enhancement: A holistic review. *Aquaculture, 519*, 734770. https://doi.org/10.1016/j.aquaculture.2019.734770.
48. Mustapha, U. F., Alhassan, A., Jiang, D., & Li, G. (2021). Sustainable

aquaculture development: A review on the roles of cloud computing, internet of things and artificial intelligence (CIA). *Reviews in Aquaculture, 13*, 2076–2091. https://doi.org/10.1111/raq.12559.

49. Niu, B. S., Li, G. Y., Peng, F., Wu, J., Zhang, L., & Li, Z. B. (2018). Survey of fish behavior analysis by computer vision. *Journal of Aquaculture Research & Development, 9*, 534–548. https://doi.org/10.4172/2155-9546.1000534.
50. O'Grady, M. J., Langton, D., & O'Hare, G. M. P. (2019). Edge computing: A tractable model for smart agriculture? *Artificial Intelligence in Agriculture, 3*, 42–51. https://doi.org/10.1016/j.aiia.2019.12.001.
51. Pau, M. B., Gabriela, A. G., José, M. V. G., Vicente, A. V., Vicente, P. P., & Víctor, E. (2018). Automatic Bluefin Tuna sizing using a stereoscopic vision system. *ICES Journal of Marine Science, 75*, 390–401. https://doi.org/10.1093/icesjms/fsx151.
52. Pedersen, S., & Wik, T. (2020). A comparison of topologies in recirculating aquaculture systems using simulation and optimization. *Aquaculture Engineering, 89*, 102059. https://doi.org/10.1016/j.aquaeng.2020.102059.
53. Peixoto, S., Soares, R., & Davis, D. A. (2020). An acoustic-based approach to evaluate the effect of different diet lengths on feeding behavior of *Litopenaeus vannamei*. *Aquaculture Engineering, 91*, 102114. https://doi.org/10.1016/j.aquaeng.2020.102114.
54. Peng, X. H., Chen, R. F., & Chen, L. (2018). Multi-parameter joint control system of aquaculture water quality based on FLC-DE. *Journal of Guangdong Ocean University, 38*, 69–75. https://doi.org/10.3969/j.issn.1673-9159.2018.06.011.
55. Peters, N. E., & Meybeck, M. (2000). Water quality degradation effects on freshwater availability: Impacts of human activities. *Water International, 25*, 185–193. https://doi.org/10.1080/02508060008686817.
56. Qiao, X., Bao, J. H., Zeng, L. H., Zou, J., & Li, D. L. (2017). An automatic active contour method for sea cucumber segmentation in natural underwater environments. *Computers and Electronics in Agriculture, 135*, 134–142. https://doi.org/10.1016/j.compag.2017.02.008.
57. Rahim, B., Mohammad, T. A., & Jan, A. (2020). Short-term water quality variable prediction using a hybrid CNN-LSTM deep learning model. *Stochastic Environmental Research and Risk Assessment, 34*, 415–433. https://doi.org/10.1007/s00477-020-01776-2.
58. Ravalli, A., Rossi, C., & Marrazza, G. (2017). Bio-inspired fish robot based on chemical sensors. *Sensors and Actuators B: Chemical, 239*, 325–329. https://doi.org/10.1016/j.snb.2016.08.030.

59. Raza, K., & Song, H. (2020). Fast and accurate fish detection design with improved YOLO-v3 model and transfer learning. *International Journal of Advanced Computer Science and Applications, 11*, 7–16. https://doi.org/10.14569/IJACSA.2020.0110202.
60. Rembold, U. (2020). *Robot Technology and Applications*. CRC Press.
61. Saberioon, M., Gholizadeh, A., Cisar, P., Pautsina, A., & Urban, J. (2016). Application of machine vision systems in aquaculture with emphasis on fish: State-of-the-art and key issues. *Reviews in Aquaculture, 9*, 1–19. https://doi.org/10.1111/raq.12143.
62. Sarma, K., Kumar, A. A., George, G., Pandian, K., Prabakaran, K., Dam Roy, S., & Srivastava, R. S. (2013). Impact of coastal pollution on biological and biochemical and nutritional status of edible oyster in Phoenix Bay Jetty and North Wandoor of Andaman. *Indian Journal of Animal Science, 83*, 89–93.
63. Shi, C., Wang, Q. B., He, X. L., Zhang, X. S., & Li, D. L. (2020). An automatic method of fish length estimation using underwater stereo system based on LabVIEW. *Computers and Electronics in Agriculture, 173*, 105419. https://doi.org/10.1016/j.compag.2020.105419.
64. Shi, W., Jie, C., Quan, Z., Li, Y., & Xu, L. (2016). Edge computing: Vision and challenges. *IEEE Internet of Things Journal, 3*(5), 637–646. https://doi.org/10.1109/JIOT.2016.2579198.
65. Sun, M., Hassan, S. G., & Li, D. L. (2016). Models for estimating feed intake in aquaculture: A review. *Computers and Electronics in Agriculture, 127*, 425–438. https://doi.org/10.1016/j.compag.2016.06.024.
66. Tai, H., Yang, Y., Liu, S., & Li, D. (2016). A review of measurement methods of dissolved oxygen in water. Berlin Heidelberg: Springer.
67. Vatsos, I. N., & Angelidis, P. (2017). Water quality and fish diseases. *Journal of the Hellenic Veterinary Medical Society, 61*, 40. https://doi.org/10.12681/jhvms.14875.
68. Viazzi, S., Hoestenberghe, S. V., Goddeeris, B. M., & Berckmans, D. (2015). Automatic mass estimation of Jade perch Scortum barcoo by computer vision. *Aquacultural Engineering, 64*, 42–48. https://doi.org/10.1016/j.aquaeng.2014.11.003.
69. Vrecko, D., Hvala, N., & Strazar, M. (2011). The application of model predictive control of ammonia nitrogen in an activated sludge process. *Water Science and Technology, 64*, 1115–1121. https://doi.org/10.2166/wst.2011.477.
70. Waleed, A., Medhat, H., Esmail, M., Osama, K., Samy, R., & Ghanim,

T. M. (2019). Automatic recognition of fish diseases in fish farms. *14th International Conference on Computer Engineering and Systems (ICCES)*, 201–206. https://doi.org/10.1109/ICCES48960.2019.9068141.

71. Wang, T., Xu, X., Wang, C., Li, Z., & Li, D. (2021). From smart farming towards unmanned farms: A new mode of agricultural production. *Agriculture, 11*, 145. https://doi.org/10.3390/agriculture11020145.
72. Wang, Y. P., Nie, Y. M., Xie, C. J., & Shuang, F. (2015). Design and research of intelligent feeding system for farmed fish based on machine vision. *Instrumentation Technology, 1*, 1–4. https://doi.org/10.19432/j.cnki.issn1006-2394.2015.01.001.
73. Wik, T., Lindén, B. T., & Wramner, P. I. (2009). Integrated dynamic aquaculture and wastewater treatment modelling for recirculating aquaculture systems. *Aquaculture, 287*, 361–370. https://doi.org/10.1016/j.aquaculture.2008.10.056.
74. Wu, T. H., Huang, Y. I., & Chen, J. M. (2015). Development of an adaptive neural-based fuzzy inference system for feeding decision-making assessment in silver perch (*Bidyanus bidyanus*) culture. *Aquaculture Engineering, 66*, 41–51. https://doi.org/10.1016/j.aquaeng.2015.02.001.
75. Xiao, R. C., Wei, Y. G., An, D., Li, D. L., Ta, X. X., Wu, Y. H., & Ren, Q. (2018). A review on the research status and development trend of equipment in water treatment processes of recirculating aquaculture systems. *Reviews in Aquaculture, 11*, 863–895. https://doi.org/10.1111/raq.12270.
76. Yang, H., Hassan, S. G., Wang, L., & Li, D. L. (2017). Fault diagnosis method for water quality monitoring and control equipment in aquaculture based on multiple SVM combined with D-S evidence theory. *Computers and Electronics in Agriculture, 141*, 96–108. https://doi.org/10.1016/j.compag.2017.05.016.
77. Yang, L., Liu, Y. Q., Yu, H. H., Fang, X. M., Song, L. H., & Li, D. L. (2021). Computer vision models in intelligent aquaculture with emphasis on fish detection and behavior analysis: A review. *Archives of Computational Methods in Engineering, 28*, 2785–2816. https://doi.org/10.1007/s11831-020-09486-2.
78. Yang, X. T., Zhang, S., Liu, J. T., Gao, Q. F., Dong, S. L., & Zhou, C. (2020). Deep learning for smart fish farming: Applications, opportunities, and challenges. *Reviews in Aquaculture, 13*, 66–90. https://doi.org/10.1111/raq.12464.
79. Yu, H. H., Yang, L., Li, D. L., & Chen, Y. Y. (2020). A hybrid intelligent soft computing method for ammonia nitrogen prediction in aquaculture.

Information Processing in Agriculture, 8, 64–74. https://doi.org/10.1016/j.inpa.2020.04.002.

80. Zhao, S. Y., Yu, J. T., Wang, Z., & Wang, Q. (2011). Multi-function self-validating water quality monitoring sensor based on relevance vector machine. *Chinese Journal of Scientific Instrument, 32*, 1690–1691. https://doi.org/10.19650/j.cnki.cjsi.2011.08.002.
81. Zhou, C., Lin, K., Xu, D. M., Chen, L., Guo, Q., Sun, C. H., & Yang, X. T. (2018). Near-infrared computer vision and neuro-fuzzy model-based feeding decision system for fish in aquaculture. *Computers and Electronics in Agriculture, 146*, 114–124. https://doi.org/10.1016/j.compag.2018.02.006.
82. Zhou, C., Xu, D. M., Chen, L., Zhang, S., Sun, C. H., Yang, X. T., & Wang, Y. B. (2019). Evaluation of fish feeding intensity in aquaculture using a convolutional neural network and machine vision. *Aquaculture, 507*, 457–465. https://doi.org/10.1016/j.aquaculture.2019.04.056.
83. Zhou, C., Xu, D. M., Lin, K., Sun, C. H., & Yang, X. T. (2017). Intelligent feeding control methods in aquaculture with an emphasis on fish: A review. *Reviews in Aquaculture, 10*, 975–993. https://doi.org/10.1111/raq.12218.
84. Zhou, C., Zhang, B. H., Lin, K., Xu, D. M., Chen, G. W., Yang, X. T., & Sun, C. H. (2017). Near-infrared imaging to quantify the feeding behavior of fish in aquaculture. *Computers and Electronics in Agriculture, 135*, 233–241. https://doi.org/10.1016/j.compag.2017.02.013.

Chapter 4

IOT In Aquaculture: Challenges and Potential Solutions

Contents

INTRODUCTION

Aquaculture accounts for nearly half of the seafood consumed by the world's ever-expanding population. There have been attempts to introduce novel technologies into the industry in order to increase profitability, particularly the Internet of Things (IoT). It is therefore important in aquaculture to evaluate and address the challenges associated with IoT. The purpose of this review is to categorize the challenges associated with implementing IoT systems in aquaculture based on the literature. This chapter examined prototyped IoT systems reported in the literature within the last five years. Furthermore, some potential solutions to the reported challenges were discussed using the literature, which can be used by future researchers. Some challenges are common to all IoT systems used in aquaculture, while others are unique to specific species or farms. Sensor probes are often difficult to maintain due to direct contact with water. The literature addressed this issue using various designs that resulted in self-cleaning mechanisms. There were three types of specific challenges: infrastructure-related, data-related, and perception-based. The primary challenges in infrastructure are the availability of electricity and internet connectivity. Energy harvesting technologies are suggested for electricity, while satellite internet or offline systems are recommended for internet connectivity. The data-related category includes issues such as data loss due to device or network errors. Studies have shown that incorporating soft methods can help resolve the majority of these issues. Perception-related challenges investigate the barriers that keep aquaculture farmers from implementing IoT systems.

4.1. IOT IN AQUACULTURE

Aquaculture is the farming of aquatic species under controlled conditions. The main factors that contribute to successful aquaculture farming are maintaining desired water quality, providing proper nutrition, promoting breeding, and preventing diseases and predators. Aquaculture production increased by at least three percent between 2011 and 2019. This is due to the increasing global population and limited wild aquatic resources. However, many challenges in aquaculture remain unresolved. One of the most significant challenges that precision aquaculture could address is maximizing yield through efficient resource utilization. Precision aquaculture aims to achieve its goal of ensuring profitability, sustainability, and environmental protection by incorporating various innovative technologies such as artificial intelligence (AI) and the Internet of Things (IoT) into aquaculture. In other words, this type of aquaculture uses cutting-edge technologies to transform traditional experience-based practices into knowledge-based practices.

Smart aquaculture, also known as intelligent aquaculture or digital aquaculture, is a concept that involves using advanced technologies and data-driven approaches to increase the efficiency, sustainability, and productivity of aquaculture operations. Smart aquaculture aims to incorporate modern technologies such as sensors, artificial intelligence, and the internet of things into the management and operation of aquaculture systems. IoT is the process of allowing various devices to communicate with one another via a network. Figure 4.1 depicts the main components of an IoT aquaculture system. It is possible to have multiple IoT architectures depending on how different components are implemented and interact.

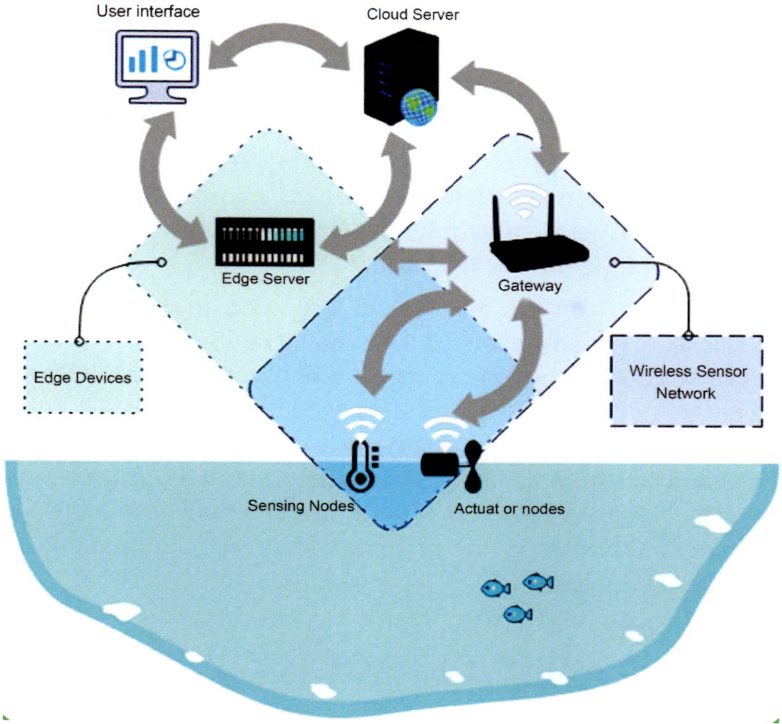

Figure 4.1. *Diagrams show how components of an IoT system interact.*

Sensing and actuating nodes perceive and respond to their surroundings. The system's nodes and gateways collectively form a Wireless Sensor Network (WSN). Edge devices are those in the IoT system that process data locally. Local edge servers, cloud servers, or both may host the IoT system's data. The final user will be able to interact with the IoT system via user interfaces.

Actuators and sensors are installed in the physical environment to either measure various water parameters or affect the environment in a specific way.

A gateway is a data transceiver that connects sensor/actuator nodes to other parts of the IoT system. IoT devices are typically connected to a Wireless Sensor Network (WSN), through which all network communication occurs. Figure 4.1 shows an example of a WSN that combines sensors, actuators, and a gateway.

Different architectures may combine cloud or edge servers to process data collected from the environment. Modern IoT devices can often perform some level of computation, and in some architectures, local servers on-premises handle a portion of the computation. In computing, edge computing refers to localized processing. This category includes all devices that perform local data processing. Figure 4.1 depicts an example of edge devices highlighted and tagged in light green. Edge computing has the advantage of allowing IoT systems to run locally without a connection to the internet. In contrast, cloud servers are typically more powerful and easier to maintain.

Nonetheless, end users can access IoT systems in a variety of ways, including via the internet or offline from a local network. Depending on the platform and software used, users may encounter a variety of user interfaces for viewing and controlling their environment.

It should be noted that aquaculture can take place in a variety of environments, both inland and off-land, including tanks, ponds, rivers, and fish pens. Similarly, different species are cultured in different aquaculture systems. The cultural environment and species composition influence the design and selection of hardware for IoT systems. An IoT system for aquaculture in a sea pen would differ from its inland counterpart. Similarly, species sensitivity influences the design and selection of hardware in Internet of Things systems. There are numerous challenges to overcome when designing and implementing IoT systems in aquaculture.

4.2. MAJOR COMPONENTS OF IOT

This section introduces various types of aquacultures before focusing on key factors that must be considered when designing IoT in aquaculture. It is followed by an introduction to the major components of IoT.

4.2.1. Aquaculture

There are three distinct types of aquaculture systems: intensive, semi-intensive, and extensive. The goal of an intensive system is to maximize culture output while using as little water as possible. As a result, a large number of aquaculture species are cultured in a small body of water. Farmers will meet all of the requirements, including providing nutrition and maintaining water quality through treatment, in order to increase productivity in the shortest amount of time. It should be noted that, due to the high costs of intensive systems, the species to be cultured are chosen based on their market value. The opposite is true for large systems that require little human intervention. These systems typically cultivate species in large bodies of water at low densities. While these systems are not expensive, the harvest yield is extremely low. Semi-intensive systems, which fall somewhere between the other two systems, treat water and provide nutrients to species in a moderate amount during the culturing phase.

IoT integration in aquaculture aims to reduce the cost of intensive systems in the aquaculture industry. It should be noted, however, that adapting IoT may cost farmers money, which is why it is critical to do so based on the unique needs of each system.

Water quality is the most important determinant factor in aquaculture, regardless of species. The majority of the proposed IoT systems in the literature focus on monitoring water quality factors such as temperature, pH, alkalinity, and nitrate level. According to a review of the literature, temperature, dissolved oxygen, and pH are the most extensively studied water quality parameters.

Each water quality factor has predefined acceptable ranges based on the species. Maintaining acceptable water quality levels is essential for achieving a successful yield from an aquaculture farm. Furthermore, maintaining optimal quality factors can reduce mortality rates while increasing yield.

4.2.1.1. Type of Species

Inhabitant factors differ depending on the aquatic species. Although some species are known for their resistance to environmental factors, others are highly sensitive. Tilapia, for example, is one of the easiest fish to culture because they are highly adaptable to changing environmental factors such as water temperature. A study of 182 shrimp farms found that the farmers' primary concern is the species' susceptibility to disease. As a precaution, farmers limited the number of visitors to their farms. Incorporating IoT in farms with sensitive species may result in reduced human interaction on the farm, lowering risks. Designing and implementing IoT systems requires an understanding of the species' resistance. More sensitive species require a system with higher accuracy and reliability than species with low resistance. Furthermore, the population density of each species has a direct impact on the design of various IoT systems. As species density rises, so will the complexity and cost of IoT systems. It has been demonstrated that proper husbandry methods can reduce stress in aquaculture. Stress can reduce the species' growth rate and weaken its immune system. Reducing stress is critical in aquaculture because it has a direct relationship with mortality, growth rate, and increases species vulnerability to disease. Biomass statistics necessitate the calculation of species weight. Traditionally, it has been done by catching and weighting species, which is a labor-intensive and stressful process for the species. Furthermore, water quality parameters like temperature and pH are identified as stressors. The use of IoT in aquaculture could reduce sources of stress by maintaining optimal water quality while also reducing human intervention. For example, using image processing to predict fish weights increased the growth rate of fish by 47%. The use of IoT could reduce human presence in the farm, lowering the risk of contamination. Furthermore, IoT could enable faster detection of contamination, resulting in earlier treatment procedures and higher survival rates.

4.2.1.2. Accessibility of the Farm

Aquaculture farming options vary, as do environmental factors, which have a direct impact on the types and designs of IoT systems. For example, access to power grids and internet connections on a farm is a critical factor influencing IoT design. Traditionally, farms are located far from power lines. In contrast, hatcheries are typically located in urban areas, with indoor tanks and access to environmental control equipment. A reliable source of electricity and internet access are also likely to be available. Some marine aquaculture farms are located in the sea, with no access to power or internet. Furthermore, some of the farms are located in areas with harsh weather. Marine aquaculture requires more maintenance and care than freshwater aquaculture because salt exists and can accumulate over time, causing corrosion to electrical equipment. Essentially, the cost of the IoT system is expected to rise with the farm's distance from the urban area where the power grid and internet connection are available. On the other hand, hatcheries are expected to work with more sensitive species than other types of farms. As a result, more reliable sensors and actuators may be required when compared to other types of farms. In other words, a variety of factors influence the equipment selection and design of the IoT system.

4.2.2. Internet of Things

In IoT systems, the architecture refers to how different components are organized into layers. The literature proposes various IoT architectures, such as three-layered, four-layered, and so on. Nonetheless, three-layered architecture is the most fundamental architecture that encompasses the core functions of IoT. The complexity of the IoT system usually determines the architecture used. It would be more beneficial to adapt more sophisticated architectures to more complex

IoT systems. The three-layered model has three layers: perception, network, and application layers. The perception layer consists of nodes that interact with the physical environment, such as sensors and actuators. The next layer is the network layer, which allows the IoT system to transmit and process data to the next layer. Finally, the application layer is the highest level that provides services to the end user. The following sections present the building blocks of the IoT system's three-layered architecture.

4.2.2.1. Perception Layer

The perception layer includes nodes, according to the three-layered architecture. A node can be either a sensor that monitors the environment or an actuator that modifies the environment. A node's other components, in addition to the sensor/actuator, include data processing, communication, and power supply subsystems. Nowadays, most sensors operate on voltage with very low current draw. In some cases, IoT systems with battery-powered sensors can provide a practical solution.

4.2.2.2. **Sensors**

Sensors are the main hardware constituent of any IoT system. Sensors empower IoT systems to gather environmental data in quantitative values. There are different sensors available in the market with different features that affect their price. Similar sensors with different qualities may have the same accuracy but different prices. For example, industrial sensors are designed to be heavily used in harsh environments while laboratory-made sensors are meant for regular usage in a controlled environment. Both industrial and laboratory build types are expected to have high accuracy. Available sensors in the market range from industrial sensors that are enjoying high accuracy and build quality to hobby sensors with reasonable accuracy and weak build quality. Usually, there is a direct relationship between price and build quality of sensors. The higher the price, the higher the accuracy and robustness. Figure 4.2 depicts the relationship between price and accuracy (°C) for different temperature sensors available in the Malaysian market. As it can be seen sensors with higher accuracy (larger radius) are more expensive. It has to be noted that many sensors in the market are priced based on additional features and therefore one may find low/similar accuracy sensors with different prices. For example, in Figure 4.2, five different prices were available for sensors with an accuracy of 0.5. Nevertheless, the aforementioned relation between price and accuracy holds as before. A general trend in the sensor market is that high-accuracy sensors are becoming more affordable over the course of time and the accuracy of the sensors in the market is more or less the same.

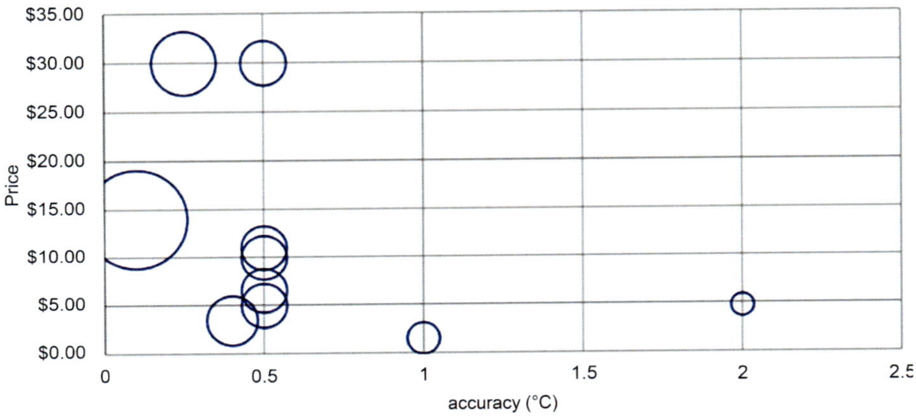

Figure 4.2. *Compares the price and accuracy of various temperature sensors available on the market. Each circle represents a sensor whose radius is equal to 1/accuracy. Supplementary materials contain related data.*

While cheap sensors are associated with lower accuracy, this does not imply that they are useless. Depending on the aquaculture environment's characteristics, a cost-benefit analysis may recommend purchasing sensors with lower build types. Instead of maintaining a sensor with a high-quality build type, it may be worthwhile to replace them with new ones on a regular basis to save money on maintenance. Nonetheless, if the goal is to reduce costs, the use of low-cost sensors may be considered depending on the type of environment and sensitivity of aquaculture species. Figure 4.2 shows a cluster of reasonably priced sensors. Temperature sensors have an error rate of around ±0.5 °C, making them suitable for most aquaculture applications. Furthermore, the use of edge computing and various algorithms has shown the ability to improve sensor accuracy. Water quality is the most important factor in aquaculture, so most IoT aquaculture projects used a variety of sensors to monitor it. These sensors measure a wide range of parameters, including dissolved oxygen (DO), pH, temperature, and water level. Other types of sensors, such as cameras, are, however, used in aquaculture. One e□ample is the use of cameras to analyze fish growth rates.

4.2.2.3. Actuators

Actuators are the means by which an IoT system can change the environment. Actuators can be used in aquaculture to activate alarms or power aerators. While the use of actuators in aquaculture increases the complexity of IoT systems, it may reduce the need for human intervention, potentially lowering labor costs as well.

Farming of sensitive species necessitates precise implementation of husbandry methods throughout the culturing period. As a result, this culture is associated with high costs and the risk of failure. These costs are primarily due to the need to maintain an optimal cultural environment, which necessitates continuous monitoring and treatment by human agents. IoT systems are capable of performing such tasks accurately without human intervention, thereby lowering associated costs. An example would be to monitor the Dissolved Oxygen (DO) level and activate the aerator if it falls below a predetermined level.

Labor costs account for the largest portion of aquaculture's operating costs. Traditional feeding is particularly labor-intensive. Furthermore, studies have shown that more than 60% of feeding material eventually converts into toxic compounds that degrade water quality. The incorporation of automated feeding tools may reduce farm operating costs in the first place. Furthermore, it can release exactly the right amount of food, reducing waste. In such a case, the actuator could be an automated dispatcher that precisely releases a predetermined amount of food or other physiochemical substances into the culturing environment. Drones can also be used to perform feeding tasks in large culturing areas.

A proposed prototype used a set of sensors to control water quality factors. In cases where DO levels fall below a predefined threshold, the proposed system can use aerators to increase them. According to reports, adding freshwater and increasing the DO level of the culture environment are critical for maintaining optimal water quality levels for farming. Their proposed system automated a water pump and aerators, allowing for freshwater addition and increased water DO levels. Similarly, a propeller was programmed to increase the DO of water if sensory measurements indicated a significant deviation from the predefined acceptable value. Another work proposed a prototype for the intelligent placement and control of an array of aerators in fish ponds. Each aerator was able to operate automatically based on the most recent readings from sensors located throughout the pond. Another study used multiple actuators such as a heater, sodium bicarbonate distributer, solenoid valve, and water pump to monitor and maintain six different water quality parameters at the desired levels. It is important to note that actuator devices in aquaculture typically require more power than sensory devices. As a result, relying on automated actuators in an IoT system is only possible if the aerator's power supply requirements are met. Integrating various intelligent methods could enable intelligent actuator activation. This is due to the collection of data from various sources and feeding it into a model that will activate the actuators. By doing so, the IoT system will be able to run more efficiently.

4.2.2.4. Network Layer

In a three-layered architecture, the network layer manages all networking between the gateway node, which serves as a transceiver, and the cloud. However, there is another type of network communication that can occur between IoT nodes, namely sensors, actuators, and gateway. The literature refers to this as the Wireless Sensor and Actuator Network (WSAN) or the Wireless Sensor Network (WSN). In this network, different nodes could communicate with one another in order to transfer data from one node to the system's gateway.

The main characteristic of IoT devices is that they are typically low-powered and designed to run on batteries. Minimal energy consumption is required to extend battery life. As a result, low power consumption becomes a deciding factor when developing networking protocols. However, in addition to power consumption, the distance between nodes and the transfer rate are the two most important factors that distinguish protocols. Figure 4.3 compares protocols based on their operating distance and transfer rate. The vertical axis represents the maximum distance that a protocol can transfer data. The higher a protocol is on the vertical axis, the longer it can transmit data. The horizontal axis depicts the transfer rate of a given protocol. The protocols on the right side of the plot can transfer more data than those on the left side of the axis.

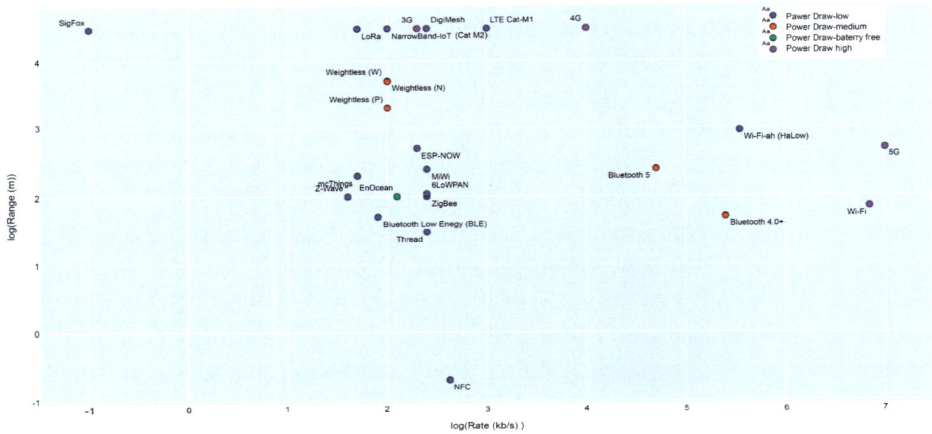

Figure 4.3. *A comparison of various communication protocols based on transmission rate and maximum operating distance. The logarithm of the actual value replaces values on both axes.*

4.2.2.5. Application Layer (Platform/Software)

IoT platforms use Application Programming Interfaces (APIs) to connect users to the core of IoT devices. Because of the inherent challenges of IoT in aquaculture, selecting the right platform requires careful consideration. The entire IoT system should be able to function even if there is no internet connection. In other words, if there is no access to the cloud, a platform should be able to function independently. All necessary decision-making and automation should continue without reliance on the internet. There must be the ability to perform the necessary critical analysis offline. This is due to the possibility of internet connectivity issues delaying data transfer to the cloud for an unknown period of time.

In an ideal IoT system, the collection of related data is critical. Analyzing data could reveal hidden insights about a farm. However, collected data may only be used for visualization or, in some cases, to compare current data to a predefined threshold.

Data can be easily stored and visualized using custom software or a cloud-based platform. The internet is an essential component of cloud solutions, despite their global accessibility and low maintenance costs. This could be difficult in aquaculture because the farm is in a remote area. Furthermore, when a cloud platform is in use, any disconnect from the cloud could have disastrous consequences. It is especially true in situations where the IoT system is intended to be used for environmental control.

Traditionally, it is assumed that nodes lack processing power, so the collected data must be stored and analyzed in a dedicated device known as a hub/gateway. Hubs can be implemented as local servers, cloud servers, or hybrids with both local and cloud servers in the topology. Platforms with local and hybrid topologies can perform even without a cloud connection, making them a viable option for aquaculture farms.

It would be prohibitively expensive to develop or customize existing open source aquaculture software platforms. It is true that the majority of the costs will be incurred during the development of the first version of any software. In other words, the second copy of a software costs almost nothing, with the exception of a reasonable maintenance fee. As a result, ordering a custom software platform would not be a viable solution for small farm businesses. A possible solution would be to share and divide costs across multiple farms. Another option for small business farmers that may be even cheaper is to subscribe to one of the available cloud-based solutions or to use an open source platform on-premises.

4.3. CHALLENGES OF IOT IN AQUACULTURE

There are two main types of challenges. The first category (Common challenges) covers general issues that may arise in most IoT systems, including aquaculture. The second category (Specific Challenges) addresses specific challenges faced by various aspects of aquaculture, such as a lack of electricity, inadequate communication coverage, a harsh environment, and insufficient technology.

4.3.1. Common Challenges

Sensors' permanent contact with polluted water reduces their accuracy over time. To address this issue, the literature suggests either regular probe maintenance, i.e., cleaning, or, in some cases, the development of special mechanisms for automatic probe cleaning.

The addition of a sensor cleaning mechanism raises the total cost of any IoT solution. However, it ensures an accurate assessment of the need for manual maintenance. The cleaning system's design must be such that sensor readings are unaffected. In an effort to clean sensors, a mechanism was developed that disperses air bubbles around sensors to prevent dirt from accumulating on their surfaces. Because of the high concentrations of oxygen molecules surrounding the probe, such solutions may produce inaccurate dissolved oxygen readings. A more effective solution would be to use ultrasonic cleaners to automate the cleaning of sensor probes. A self-cleaning mechanism was designed to clean the sensor probe after each measurement. This was accomplished by creating a specialized robotic arm that automatically performs the measurement and rinses the sensor probe. If the sensor is not to be used for an extended period of time, the proposed system submerges it in a protective solution.

4.3.2. Specific Challenges

Certain challenges arise more frequently in different farming setups depending on the species and farm's accessibility. Table 4.1 lists the challenges to implementing IoT in aquaculture systems in two dimensions. The rows represent the farm's access to basic facilities, while the columns represent the species under cultivation. A farm that does not have access to electricity or the internet is considered rural. Farms in urban areas have access to both of these amenities. A farm with basic facilities will face fewer challenges than a farm in a rural area. Similarly, sensitive species expect greater dependability and quality in design, hardware, and software. In contrast, less sensitive species could benefit from simpler and more cost-effective IoT systems. Such Categorization resulted in four different scenarios, with each farm falling into one of them. There are various challenges associated with each scenario, which are classified into three categories: infrastructure-related challenges, data-related challenges, and perception challenges. (Rastegari et al., 2023).

Table 4.1. *Illustrates the Various Types of Challenges that an IoT System may Face*

Empty Cell	Empty Cell	**Type of Species**	
Empty Cell	Empty Cell	**Sensitive**	**Insensitive**
Accessibility of the Farm	Rural	• Infrastructure Related • Data Related • Perception Related	• Infrastructure Related • Perception Related
Empty Cell	Urban	• Data Quality • Perception Related	• Perception Related

Some challenges are more likely to occur based on the species and the farm's proximity to facilities. Culturists should consider the costs associated with Table 4.1. The number of challenges increases the cost of implementing the IoT system. It is due to the numerous challenges that must be addressed. As can be seen, the majority of the challenges are faced by farms with sensitive species located in rural areas. One of the approaches to addressing the challenges could be as follows. To reduce the number of challenges, start by changing the species or moving the farm. Second, confront them by addressing the challenge with various methodologies proposed in the literature, as discussed below. To this end, a thorough cost-benefit analysis is required before deciding on one of the approaches mentioned above. For established farms, the first approach may be impractical. In contrast, for future farms, it may be prudent to consider the location and type of species in a way that reduces the number of obstacles. Each category of challenge will be discussed further below, along with research findings on how such challenges have been addressed in the literature.

4.3.3. Infrastructure Related Challenges

The availability of electricity and internet connectivity on the farm are significant infrastructure challenges that can influence the design of IoT systems. Each of these topics will be further examined in the subsequent sections.

4.3.3.1. Electricity

IoT systems rely on electricity for their operation. The absence of electricity in rural regions or marine aquaculture, along with the inconsistent nature of electricity in urban areas, have been recognized as obstacles that require attention. The majority of Internet of Things (IoT) devices are specifically engineered to function with minimal power consumption, enabling them to operate for prolonged durations using battery power. Nevertheless, potential actuators generally have a high-power consumption, making it impractical to rely solely on a battery without access to a recharging facility. Nevertheless, the advent of recently developed betavoltaic batteries, which can last for several decades to centuries, could potentially eliminate this requirement in the foreseeable future. The utilization of solar panels and photovoltaic batteries has demonstrated its effectiveness as a practical solution to this issue. Nevertheless, the dependable functioning of other solar-powered equipment necessitates meticulous design. For instance, the dimensions of the pond have an impact on the consistent operation of a solar photovoltaic and battery-powered aerator for a fish farm. It is important to mention that although the installation of solar panels and batteries would raise the farm's total expenses, it has been demonstrated to be more economically efficient than using gasoline or relying on electricity from the power grid.

The active research area of optimizing energy consumption aims to enhance the longevity of battery-operated devices. Empirical evidence has shown that decreasing the frequency of data transmissions across the network can conserve energy while maintaining a high level of data integrity.

Most of the suggested remedies in the literature depend on real-time monitoring of water quality. Continuous monitoring of water quality requires a significant amount of power. Therefore, in order to ma□imize the lifespan of the battery, it is important to optimize power consumption by minimizing the duration of sampling periods as much as possible. Utilizing existing technology. IoT nodes have the capability to enter sleep mode, resulting in a substantial reduction in power consumption. Implementing periodic sampling may decrease the quantity of readings, but by integrating machine learning techniques, the predictions generated are typically similar to the actual readings.

Connectivity to the internet is a vital necessity for IoT systems. Rural farms may face a substantial challenge as a result. The availability of internet connectivity enables the Internet of Things (IoT) system to transmit data to the cloud, facilitating remote analysis and monitoring. Multiple networking protocols can be utilized to transmit data between the farm and the internet. Nevertheless, it is important to acknowledge that long-range networking protocols are specifically designed for the transmission of data at a low rate. Consequently, although the majority of the requirements for an IoT aquafarm will be fulfilled, specific requirements like video data transfer will not be satisfied by most of these technologies. A comparative study was undertaken to evaluate various low-power lossy networks (LLNs) in the context of Internet of Things (IoT) for the purpose of video transmission. Their assessment indicated that most current protocols lack the capability to transmit video. According to the authors, video cannot be transmitted over these networks due to their inherent characteristics. Nevertheless, alternative protocols like CoAP have the capability to transmit data at a fast rate, including video content.

Recent technological advancements unequivocally prove that achieving global internet access is now a tangible reality. The technology of reliable high-speed internet has been accessible in certain regions of the world for a number of years and is anticipated to be extended to additional areas in the near future. Nevertheless, like any nascent technology, embracing it at an early stage will incur significant e□penses. Sharing can be considered as a viable approach for small farms located in rural areas to decrease the expenses associated with adopting new practices.

Edge computing has enabled the execution of the majority of computations at the location where they are needed, in recent times. Edge computing has the potential to enhance the quality of service (QoS), reduce response time, and bolster security. Moreover, when there is no need for extra data from external sources to perform computations, the IoT system becomes less dependent on the cloud. Put simply, the utilization of edge computing in an IoT system enables the system to operate independently without depending on cloud resources. The viability of implementing this architecture for a farm would rely on the farm's capacity to establish an internal network. It is important to mention that edge computing does not separate an IoT system. Put simply, edge computing decreases the IoT system's dependence on high-performance cloud computing machines accessible via the internet. It facilitates decentralized decision-making in an Internet of Things system. FOG computing, or a combination of edge and fog computing, could be advantageous for farms that have limited network accessibility. By employing fog computing, it is possible to apply data filtration, thereby diminishing the volume of data that necessitates transmission over the internet.

A frequent problem, especially for offshore farms, is the occurrence of cable fractures in underwater sensor networks. Optimizing the wireless network and suggesting a combination of wireless and wired topology have demonstrated their efficacy as solutions.

4.3.4. Data-Related Challenges

The accuracy and reliability of collected data are crucial as the primary objective of an IoT system is to facilitate data-informed decision making. Data quality is influenced by various factors throughout the process, starting from the data source and extending to the quality of sensors, which plays a crucial role in gathering data of high quality.

4.3.4.1. Hardware Malfunction

Various vendors provide diverse iterations of the identical sensor, designed to cater to a wide range of environments, spanning from recreational to industrial applications. Moreover, various sensors possess distinct characteristics, typically quantified in terms of their error rate. As the error decreases, the accuracy (quality) increases. The precision of the sensors is directly correlated with their cost. Figure 4.2 presents a comparison of the pricing and accuracy of different temperature sensors. It is evident that there is a clear relationship between the cost and precision of a temperature sensor.

Another factor contributing to data inaccuracy is the necessity for periodic maintenance of certain sensors, which may involve adhering to the manufacturer's recommended procedures or replacing consumable components. pH sensors necessitate regular recalibration. Alternatively, as mentioned earlier, the majority of sensors necessitate regular cleaning. This phenomenon can arise due to various factors, one of which is biofouling, leading to a deviation in the accuracy of the sensor. A technique was suggested for automatically purging the sensor head with compressed air using an air pump. However, this approach necessitates energy to operate the air pump and is not suitable for all settings. Exposure to water and humidity leads to corrosion in sensors and electronic components, requiring frequent maintenance.

Although more expensive sensors offer superior accuracy, the probability of receiving inaccurate data gradually rises. Therefore, the implementation of the following techniques to rectify sensor readings, along with the execution of maintenance protocols, has been demonstrated to be successful.

- Utilizing the data fusion technique to enhance data integrity while simultaneously decreasing the frequency of data transmission across the network, thereby conserving energy.
- Employing anomaly detection and other sophisticated algorithms to improve and forecast sensor values.

A network of temperature sensors with multiple depth levels was established by deploying 15 marine buoys at different locations across the sea. The proposed system comprised of two software modules: the accumulation system and the visualization system. The sensor nodes transmit temperature data to a mail server via email. The data of each node will be stored in a distinct file in te□t format, which will be utilized subsequently for visualization. Interpolation was employed to compute the temperature for sensor data that was not available, as well as to forecast the temperature value between sensors.

4.3.4.2. Connection Failure

Lossy networks impede the seamless gathering of data. The majority of Internet of Things (IoT) platforms store the gathered data in the form of time series. The loss of data at various intervals can significantly affect the overall computation. Packet loss in networks is a widely studied subject that refers to the failure of data packets to successfully transition due to factors such as low received signal strength indicator (RSSI) values and signal-to-noise ratios (SNR). Repeating the identical reading multiple times can decrease the likelihood of data loss. To tackle this problem, one could consider implementing different strategies, such as interpolation or other statistical techniques, improving the reliability of radio and antenna configuration, or even setting up an IoT platform on a local server.

The topological configuration of the IoT system can also help mitigate network-related problems. A proposition was made for a water quality monitoring system that operates using agents. Every sensor node has the option to establish a direct connection with the gateway, or alternatively, transmit data to the gateway through other nodes within the network.

4.3.5. Perception-Related Challenges

According to an empirical study, the majority of small business farmers believe that IoT is ineffective in aquaculture. The main reason for this finding was a lack of confrontation among farmers with successful exemplary farms. While large industrial scale companies reap the benefits of IoT in aquaculture, small businesses continue to resist its adoption.

Many authors suggested that policymakers establish pilot farms equipped with IoT facilities and empirically demonstrate their effectiveness to farmers in order to facilitate adoption.

The cost of IoT systems has also served as a barrier to adoption for small-scale aquaculture businesses. While it is recommended to use industrial grade sensors whenever possible, it has been demonstrated that low-cost sensors can also be used to implement practical and reliable IoT systems.

While technical issues such as power outages and internet access must be addressed, perceptual issues regarding the utility of IoT must be addressed even sooner. One of the primary reasons for small aquaculture businesses' failure to adopt new technologies, including IoT, was determined to be a lack of economic power. Because implementation costs increase linearly with system size, the amount of investment for farmers with small farms may be considered low, and thus providing an economic incentive, such as an IoT adaptation loan for small businesses, may persuade farmers to adopt new technologies. Furthermore, another major factor identified as a lack of confidence in technological farms. Technology adoption may result in more efficient products, and literature suggests that governments should support small businesses. It has been demonstrated that the adoption of IoT can be accomplished using low-cost solutions. In other words, expensive industrial-grade implementation isn't the only viable aquaculture option. It is especially beneficial to small-scale farmers, who are typically disadvantaged due to a lack of economic power. The economical design of the IoT system may also reduce costs without sacrificing quality. For example, combining the gateway and a node eliminated the need for a specific gateway node, resulting in a more cost-effective system.

4.4. FUTURE TRENDS

It is difficult to predict future trends with certainty because the Internet of Things (IoT) is constantly evolving. Some potential future trends that may help tackling the challenges of IoT for aquaculture are discussed in terms of the type of challenge, which are classified as follows:

4.4.1. Common Challenges

Antifouling method development is an active research field with the potential for use in aquaculture sensors. These methods place a greater emphasis on the use of chemical and nanocomposite technology to reduce biofilm accumulation on the surfaces of sensors and other equipment in direct contact with water. It is worth noting that few studies appear to have focused on the incorporation of self-cleaning methods for sensors in aquaculture. As a result, applying current technologies would be beneficial in this regard.

4.4.2. Specific Challenges

The specific challenges in aquaculture vary based on factors such as the species being farmed and the location of the farm. Therefore, it is unlikely that all aquaculture IoT systems will encounter the challenges described in this category. However, the introduction of energy harvesting and renewable energy technologies, such as solar, wind, and wave power, is anticipated to mitigate some of the difficulties related to energy and electricity in rural farming areas. Although floating solar panels and wind turbines are currently viable and developing technologies, the utilization of emerging energy sources like hydrogen has demonstrated promising potential.

The progress in battery manufacturing industry could potentially affect the difficulties faced by IoT systems in aquaculture. Although numerous IoT devices and sensors operate on low power, the utilization of long-lasting batteries like betavoltaic batteries could potentially diminish the energy requirements, particularly in rural regions.

Regarding the previously mentioned data-related challenges, the utilization of diverse software methodologies has demonstrated success. In addition, data fusion and sensor fusion are areas of ongoing research that can be utilized to tackle network and data errors, while simultaneously enhancing data precision and dependability.

4.5. CONCLUSION

The utilization of Internet of Things (IoT) is extensive across virtually every sector. Aquaculture, a subfield of agriculture, is no different. Although the Internet of Things presents various difficulties overall, there are certain challenges that are unique to the field of aquaculture. This chapter focuses on the documented obstacles encountered during the deployment of Internet of Things (IoT) systems in the field of aquaculture. In addition, potential resolutions to the documented difficulties were deliberated.

Immersing an IoT device in water for a prolonged duration is commonly believed to cause a significant accumulation on its surface. Consequently, the system may experience malfunctions. It is particularly crucial for sensory nodes that have the potential to generate imprecise measurements. The literature suggests a range of proposed solutions, including recommending regular maintenance of these devices and designing different self-cleaning mechanisms.

The main challenges related to infrastructure primarily revolve around the provision of electricity and the availability of internet connectivity. The utilization of solar energy for harvesting power is the predominant technology employed to energize a wide range of Internet of Things (IoT) systems. The utilization of edge computing and the capacity to deploy an Internet of Things (IoT) system that is independent of cloud reliance enables farms to function efficiently. In such instances, remote access may be restricted to solely monitoring essential data.

Data-related challenges emerge due to sensor or network errors, prompting the utilization of diverse soft techniques as recommended in the literature.

Perception-related challenges pertain to the issues that arise when farmers need to adapt to IoT systems. This is particularly applicable to the small business aquaculture model, and it is imperative for policymakers and other institutions with social responsibility to acknowledge and tackle this issue.

REFERENCES

1. Adhikari, S., Keshav, C. A., Barlaya, G., Rathod, R., Mandal, R. N., Ikmail, S., et al. (2018). Adaptation and mitigation strategies of climate change impact in freshwater aquaculture in some states of India. *Journal of Fisheries, 12*, 016–021.
2. Ahmed, N. (2013). Linking prawn and shrimp farming towards a green economy in Bangladesh: Confronting climate change. *Ocean and Coastal Management, 75*, 33–42. https://doi.org/10.1016/j.ocecoaman.2013.01.002.
3. Ahmed, N., Thompson, S., & Glaser, M. (2019). Global aquaculture productivity, environmental sustainability, and climate change adaptability. *Environmental Management, 63*, 159. https://doi.org/10.1007/s00267-018-1117-3.
4. Akegbejo-Samsons, Y. (2009). Climate change impacts on fisheries production in land-water interface. *IOP Conference Series: Earth and Environmental Science, 6*, 292022. https://doi.org/10.1088/1755-1307/6/29/292022.
5. Asiedu, B., Malcolm, D., & Iddrisu, S. (2019). Assessing the economic impact of climate change in the small-scale aquaculture industry of Ghana, West Africa. *AAS Open Research, 1*, 26. https://doi.org/10.12688/aasopenres.12911.2.
6. AskarySary, A., Velayatzadeh, M., & KarimiSary, V. (2012). Proximate composition of farmed fish, Oncorhynchus mykiss and Cyprinus carpio from Iran. *Advances in Environmental Biology, 6*, 2841–2845.
7. Bahri, T., Barange, M., & Moustahfid, H. (2018). Chapter 1: Climate change and aquatic systems. In M. Barange, T. Bahri, M. C. M. Beveridge, K. L. Cochrane, S. Funge-Smith, & F. Poulaine (Eds.), *Impacts of Climate Change on Fisheries and Aquaculture, Synthesis of Current Knowledge, Adaptation and Mitigation Options* (pp. 1–17). Rome: FAO.
8. Baker, S. M., Baker, P., Heuberger, D., & Sturmer, L. (2005). Short-term effects of rapid salinity reduction in seed clams (Mercenaria, Mercenaria). *Journal of Shellfish Research, 24*, 29–33. https://doi.org/10.2983/0730-8000(2005)24[29: SEORSR]2.0.CO;2.
9. Baker, S., Hoover, E., & Sturmer, L. (2013). The role of salinity in hard clam aquaculture. Gainesville, FL: University of Florida.
10. Barange, M. (2019). Avoiding misinterpretation of climate change projections of fish catches, food for thought. *ICES Journal of Marine Science, 76*, 1390–1392. https://doi.org/10.1093/icesjms/fsz061.

11. Barange, M., Bahri, T., Beveridge, M. C. M., Cochrane, A. L., Funge-Smith, S., & Paulain, F. (2018). *Impacts of Climate Change on Fisheries and Aquaculture, Synthesis of Current Knowledge, Adaptation and Mitigation Options*. Rome: FAO.
12. Basti, L., Endo, M., & Segawa, S. (2011). Physiological, pathological, and defense alterations in Manila clams (short-neck clams), *Ruditapes philippinarum*, induced by *Heterocapsa circularisquama*. *Journal of Shellfish Research, 30*, 829–844. https://doi.org/10.2983/035.030.0324.
13. Basti, L., Nagai, K., Segawa, S., Tanaka, Y., Toshiyuki Suzuki, T., & Nagai, S. (2019). Harmful algal blooms and shellfish aquaculture in changing environment. *Bulletin of the Japanese Fisheries Research and Education Agency, 49*, 73–79.
14. Beach, R. H., & Viator, C. L. (2008). The economics of aquaculture insurance: An overview of the U.S. pilot insurance program for cultivated clams. *Aquaculture Economics and Management, 12*, 25–38. https://doi.org/10.1080/13657300801959613.
15. Belfer, E., Ford, J. D., & Maillet, M. (2017). Representation of Indigenous peoples in climate change reporting. *Climatic Change, 145*, 57. https://doi.org/10.1007/s10584-017-2076-z.
16. Bell, J. D., Ganachaud, A., Gehrke, P. C., Griffiths, S. P., Hobday, A. J., Hoegh-Guldberg, O., et al. (2013). Mi□ed responses of tropical Pacific fisheries and aquaculture to climate change. *Nature Climate Change, 3*, 591–599. https://doi.org/10.1038/nclimate1838.
17. Bell, J., Batty, M., Ganachaud, A., Gehrke, P., Hobday, A., Hoegh-Guldberg, O., et al. (2010). Preliminary assessment of the effects of climate change on fisheries and aquaculture in the Pacific. In R. Gillett (Ed.), *Fisheries in the Economies of the Pacific Island Countries and Territories* (pp. 451–469). Manila: Asian Development Bank.
18. Beveridge, M. C. M., Dabbadie, L., Soto, D., Ross, L. G., Bueno, P. B., & Aguilar-Manjarrez, J. (2018). Chapter 22: Climate change and aquaculture: Interactions with fisheries and agriculture. Rome: FAO.
19. Black, K., & Hughes, A. (2017). Future of the sea: Trends in aquaculture, foresight–Future of the sea evidence review foresight. Edinburgh: Government Office for Science.
20. Blanchard, J. L., Watson, R. A., Fulton, E. A., Cottrell, R. S., Nash, K. L., Bryndum-Buchholz, A., et al. (2017). Linked sustainability challenges and trade-offs among fisheries, aquaculture and agriculture. *Nature Ecology & Evolution, 1*, 1240–1249. https://doi.org/10.1038/s41559-017-0258-8.

21. Bradley, M., Putten, I., & Sheaves, M. (2015). The pace and progress of adaptation: Marine climate change preparedness in Australia's coastal communities. *Marine Policy, 53*, 13–20. https://doi.org/10.1016/j.marpol.2014.11.004.
22. Brander, K. M. (2007). Global fish production and climate change. *Proceedings of the National Academy of Sciences of the United States of America, 104*, 19709–19714. https://doi.org/10.1073/pnas.0702059104.
23. Brodie, J., Williamson, C. J., Smale, D. A., Kamenos, N. A., Mieszkowska, N., Santos, R., et al. (2014). The future of the northeast Atlantic benthic flora in a high CO2 world. *Ecology and Evolution, 4*(14), 2787–2798. https://doi.org/10.1002/ece3.1105.
24. Bueno, P. B., & Soto, D. (2017). Adaptation strategies of the aquaculture sector to the impacts of climate change. Rome: FAO.
25. Canadian Institute for Climate Studies. (2000). Aquaculture and climate change in Canada: A discussion paper. Ottawa, ON: Canadian Institute for Climate Studies.
26. Chan, F. T., Stanislawczyk, K., Sneekes, A. C., Dvoretsky, A., Gollasch, S., Minchin, D., et al. (2019). Climate change opens new frontiers for marine species in the Arctic: Current trends and future invasion risks. *Global Change Biology, 25*(1), 25–38. https://doi.org/10.1111/gcb.14469.
27. Cheung, W. W. L., Watson, R., & Pauly, D. (2013). Signature of ocean warming in global fisheries catch. *Nature, 497*(7449), 365–369. https://doi.org/10.1038/nature12156.
28. Chiaramonte, L., Munson, D., & Trushenski, J. (2016). Climate change and considerations for fish health and fish health professionals. Fish Health Section, Fisheries. Available online at: www.fisheries.org (accessed on 3 August 2024).
29. Chung, I. K., Oak, J. H., Lee, J. A., Shin, J. A., Kim, J. G., & Park, K. S. (2013). Installing kelp forest/seaweed beds for mitigation and adaptation against global warming: Korean project overview. *ICES Journal of Marine Science, 70*(5), 1038–1044. https://doi.org/10.1093/icesjms/fss206.
30. Chung, I. K., Sondak, C. F. A., & Beardall, J. (2017). The future of seaweed aquaculture in a rapidly changing world. *European Journal of Phycology, 52*(4), 495–505. https://doi.org/10.1080/09670262.2017.1359678.
31. Clements, J. S., & Chopin, T. (2016). Ocean acidification and marine aquaculture in North America: Potential impacts and mitigation strategies. *Reviews in Aquaculture, 9*(4), 326–341. https://doi.org/10.1111/raq.12140.
32. Cochrane, K., De Young, C., Soto, D., & Bahri, D. T. (2009). Climate

change implications for fisheries and aquaculture: Overview of current scientific knowledge. FAO Fisheries and Aquaculture Technical Paper No. 530. Rome: FAO.

33. Collins, C., Bresnan, E., Brown, L., Falconer, L., Guilder, J., Jones, L., et al. (2020). Impacts of climate change on aquaculture. *MCCIP Science Review 2020*, 482–520. https://doi.org/10.14465/2020.arc21.aqu.
34. Cook, D., & Zolnikov, T. R. (2019). Antarctica. In T. R. Zolnikov (Ed.), *Global Adaptation and Resilience to Climate Change* (pp. 31–49). Palgrave Studies in Climate Resilient Societies. Bern; Cham: Palgrave Pivot. https://doi.org/10.1007/978-3-030-01213-7_1.
35. Cooley, S. R., Lucey, N., Kite-Powell, H., & Doney, S. C. (2012). Nutrition and income from molluscs today imply vulnerability to ocean acidification tomorrow. *Fish and Fisheries, 13*(2), 182–215. https://doi.org/10.1111/j.1467-2979.2011.00424.x.
36. Cornwall, C. E., Hepburn, C. D., Pritchard, D., Currie, K. I., McGraw, C. M., Hunter, K. A., et al. (2012). Carbon use strategies in macroalgae: Differential responses to lowered pH and implications for ocean acidification. *Journal of Phycology, 48*(1), 137–144. https://doi.org/10.1111/j.1529-8817.2011.01085.x.
37. Dabbadie, L., Aguilar-Manjarrez, J. J., Beveridge, M. C. M., Bueno, P. B., Ross, L. G., & Soto, D. (2018). Chapter 20: Effects of climate change on aquaculture: Drivers, impacts, and policies. Rome: FAO.
38. Daw, T., Adger, W. N., Brown, K., & Badjeck, M. C. (2009). Climate change and capture fisheries: Potential impacts, adaptation, and mitigation. In K. Cochrane, C. De Young, D. Soto, & T. Bahri (Eds.), *Climate Change Implications for Fisheries and Aquaculture: Overview of Current Scientific Knowledge* (pp. 107–150). FAO Fisheries and Aquaculture Technical Paper No. 530. Rome: FAO.
39. De Silva, S. S., & Soto, D. (2009). Climate change and aquaculture: Potential impacts, adaptation, and mitigation. In K. Cochrane, C. De Young, D. Soto, & T. Bahri (Eds.), *Climate Change Implications for Fisheries and Aquaculture: Overview of Current Scientific Knowledge* (pp. 151–212). FAO Fisheries and Aquaculture Technical Paper No. 530. Rome: FAO.
40. De Young, C., Soto, D., Bahri, T., & Brown, D. (2012). Building resilience for adaptation to climate change in the fisheries and aquaculture sector. In A. Meybeck, J. Lankoski, S. Redfern, N. Azzu, & V. Gitz (Eds.), *Proceedings of a Joint FAO/OECD Workshop*. Rome: FAO.
41. Doney, S. C., Ruckelshaus, M., Duffy, J. E., Barry, J. P., Chan, F., English,

C. A., et al. (2012). Climate change impacts on marine ecosystems. *Annual Review of Marine Science, 4*, 11–37. https://doi.org/10.1146/annurev-marine-041911-111611.

42. Doubleday, Z. A., Clarke, S. M., Li, X., Pecl, G. T., Ward, T. M., Battaglene, S., et al. (2013). Assessing the risk of climate change to aquaculture: A case study from south-east Australia. *Aquaculture Environment Interactions, 3*(2), 163–175. https://doi.org/10.3354/aei00058.
43. Duarte, C. M., Wu, J., Xiao, X., Bruhn, A., & Krause-Jensen, D. (2017). Can seaweed farming play a role in climate change mitigation and adaptation? *Frontiers in Marine Science, 4*, 100. https://doi.org/10.3389/fmars.2017.00100.
44. Edwards, M., & Richardson, A. J. (2004). Impact of climate change on marine pelagic phenology and trophic mismatch. *Nature, 430*(7002), 881–884. https://doi.org/10.1038/nature02808.
45. Edwards, M., Johns, D. G., Leterme, S. C., Svendsen, E., & Richardson, A. J. (2006). Regional climate change and harmful algal blooms in the NE Atlantic. *Limnology and Oceanography, 51*(2), 820–829. https://doi.org/10.4319/lo.2006.51.2.0820.
46. Ellis, R. P., Urbina, M. A., & Wilson, R. W. (2017). Lessons from two high CO2 worlds–future oceans and intensive aquaculture. *Global Change Biology, 23*(6), 2141–2148. https://doi.org/10.1111/gcb.13515.
47. Elum, Z. A., Modise, D. M., & Marr, A. (2017). Farmer's perception of climate change and responsive strategies in three selected provinces of South Africa. *Climate Risk Management, 16*, 246–257. https://doi.org/10.1016/j.crm.2016.11.001.
48. Engle, C. R. (2019). Bringing aquaculture sustainability down to earth. *Journal of the World Aquaculture Society, 50*(2), 246–248. https://doi.org/10.1111/jwas.12609.
49. FAO. (2018). *The State of World Fisheries and Aquaculture 2018: Contributing to Food Security and Nutrition for All*. Rome: FAO.
50. FAO. (2020). *The State of World Fisheries and Aquaculture 2020: Sustainability in Action*. Rome: FAO.
51. Farrell, H., Seebacher, F., O'Connor, W., Zammit, A., Harwood, D. T., & Murray, S. (2015). Warmer temperature acclimation impacts metabolism of paralytic shellfish to□ins from *Alexandrium minutum* in commercial oysters. *Global Change Biology, 21*(9), 3402–3413. https://doi.org/10.1111/gcb.12952.
52. Ficke, A. D., Myrick, C. A., & Hansen, L. J. (2007). Potential impacts of

global climate change on freshwater fisheries. *Reviews in Fish Biology and Fisheries, 17*(4), 581–613. https://doi.org/10.1007/s11160-007-9059-5.

53. Fleming, A., Hobday, A. J., Farmery, A., van Putten, E. I., Pecl, G. T., Green, B. S., et al. (2014). Climate change risks and adaptation options across Australian seafood supply chains: A preliminary assessment. *Climate Risk Management, 1*, 39–50. https://doi.org/10.1016/j.crm.2013.12.003.
54. Franke, A., & Clemmesen, C. (2011). Effect of ocean acidification on early life stages of Atlantic herring (*Clupea harengus* L.). *Biogeosciences, 8*(12), 3697–3707. https://doi.org/10.5194/bg-8-3697-2011.
55. Freeman, E. O. (2017). Impact of climate change on aquaculture and fisheries in Nigeria: A review. *International Journal of Multidisciplinary Research and Development, 4*(1), 53–59.
56. Froehlich, H. E., Gentry, R. R., & Halpern, B. S. (2017). Conservation aquaculture: Shifting the narrative and paradigm of aquaculture's role in resource management. *Biological Conservation, 215*, 162–168. https://doi.org/10.1016/j.biocon.2017.09.012.
57. Froehlich, H. E., Gentry, R. R., & Halpern, B. S. (2018). Global change in marine aquaculture production potential under climate change. *Nature Ecology & Evolution, 2*(11), 1745–1750. https://doi.org/10.1038/s41559-018-0669-1.
58. Frommel, A. Y., Maneja, R., Lowe, D., Malzahn, A. M., Geffen, A. J., Folkvord, A., et al. (2012). Severe tissue damage in Atlantic cod larvae under increasing ocean acidification. *Nature Climate Change, 2*(1), 42–46. https://doi.org/10.1038/nclimate1324.
59. Frommel, A. Y., Maneja, R., Lowe, D., Pascoe, C. K., Geffen, A. J., Folkvord, A., et al. (2014). Organ damage in Atlantic herring larvae as a result of ocean acidification. *Ecological Applications, 24*(5), 1131–1143. https://doi.org/10.1890/13-0297.1.
60. Frusher, S. D., Hobday, A. J., Jennings, S. M., Crighton, C., De Silva, S., Pecl, G. T., et al. (2013). The short history of a marine hotspot—From anecdote to adaptation in south-east Australia. *Reviews in Fish Biology and Fisheries, 24*(4), 593–611. https://doi.org/10.1007/s11160-013-9325-7.
61. Galbraith, E. D., Carozza, D. A., & Bianchi, D. (2017). A coupled human-Earth model perspective on long-term trends in the global marine fishery. *Nature Communications, 8*, 14884. https://doi.org/10.1038/ncomms14884.
62. Gao, Y., Yu, G., Yang, T., Jia, Y., He, N., & Zhuang, J. (2016). New insight into global blue carbon estimation under human activity in land-

sea interaction areas: A case study of China. *Earth-Science Reviews, 159*, 36–46. https://doi.org/10.1016/j.earscirev.2016.05.003.

63. Gazeau, F., Alliouane, S., Bock, C., Bramanti, L., Gentille, M., Hirse, T., et al. (2014). Impact of ocean acidification and warming on the Mediterranean mussel (*Mytilus galloprovincialis*). *Frontiers in Marine Science, 1*, 62. https://doi.org/10.3389/fmars.2014.00062.

64. Gubbins, M., Bricknell, I., & Service, M. (2013). Impacts of climate change on aquaculture. *MCCIP Science Review 2013*, 318–327. https://doi.org/10.14465/2013.arc33.318-327.

65. Guyondet, T., Comeau, L. A., Bacher, C., Grant, J., Rosland, R., Sonier, R., et al. (2018). Climate change influences carrying capacity in a coastal embayment dedicated to shellfish aquaculture. *Estuaries and Coasts, 38*(5), 1593–1618. https://doi.org/10.1007/s12237-014-9899-x.

66. Haberkorn, H., Lambert, C., Le Goïc, N., Guéguen, M., Moal, J., Palacios, E., et al. (2010). Effects of *Alexandrium minutum* exposure upon physiological and hematological variables of diploid and triploid oysters, *Crassostrea gigas*. *Aquatic Toxicology, 97*(1), 96–108. https://doi.org/10.1016/j.aquatox.2009.12.006.

67. Hambal, H., Mohd. Akhir, A., & Saniah, Y. (1994). Environmental issues on aquaculture development in Malaysia. *Working Papers of the Fisheries Research Institute, Department of Fisheries, Kuala Lumpur, Malaysia*.

68. Hamdan, R., Kari, F., Othman, A., & Samsi, S. M. (2012). Climate change, socio-economic and production linkages in East Malaysia aquaculture sector. In *2012 International Conference on Future Environment and Energy IPCBEE* (Vol. 28). Singapore: IACSIT Press.

69. Hamdan, R., Othman, A., & Kari, F. (2015). Climate change effects on aquaculture production performance in Malaysia: An environmental performance analysis. *International Journal of Business and Society, 16*(3), 364–385. https://doi.org/10.33736/ijbs.573.2015.

70. Handisyde, N. T., Ross, L. G., Badjeck, M. C., & Allison, E. H. (2006). *The Effects of Climate Change on World Aquaculture: A Global Perspective*. Final Technical Report. Stirling: DFID Aquaculture and Fish Genetics Research Programme, Stirling Institute of Aquaculture.

71. Handisyde, N., Telfer, T. C., & Ross, L. G. (2017). Vulnerability of aquaculture-related livelihoods to changing climate at the global scale. *Fish and Fisheries, 18*(3), 466–488. https://doi.org/10.1111/faf.12186.

72. Hardy, H. (2010). Utilization of plant proteins in fish diets: Effects of global demand and supplies of fish meal. *Aquaculture Research, 41*(6), 770–776. https://doi.org/10.1111/j.1365-2109.2009.02349.x.

73. Harvey, B., Soto, D., Carolsfeld, J., Beveridge, M., & Bartley, D. M. (2017). *Planning for Aquaculture Diversification: The Importance of Climate Change and Other Drivers*. FAO Technical Workshop, 23–25 June 2016. FAO Fisheries and Aquaculture Proceedings No. 47. Rome: FAO.
74. Hégaret, H., Brokordt, K. B., Gaymer, C. F., Lohrmann, K. B., García, C., & Varela, D. (2012). Effects of the to□ic dinoflagellate *Alexandrium catenella* on histopathological and escape responses of the Northern scallop *Argopecten purpuratus*. *Harmful Algae, 18*, 74–83. https://doi.org/10.1016/j.hal.2012.04.006.
75. Hinder, S. L., Hays, G. C., Edwards, M., Roberts, E. C., Walne, A. W., & Gravenor, M. B. (2012). Changes in marine dinoflagellate and diatom abundance under climate change. *Nature Climate Change, 2*(4), 271–275. https://doi.org/10.1038/nclimate1388.
76. Hoegh-Guldberg, O., Mumby, P. J., Hooten, A. J., Steneck, R. S., Greenfield, P., Gomez, E., et al. (2007). Coral reefs under rapid climate change and ocean acidification. *Science, 318*(5857), 1737–1742. https://doi.org/10.1126/science.1152509.
77. Holmyard, N. (2014). Climate change: Implications for fisheries and aquaculture. In J. Benn (Ed.), *Key Findings from the Intergovernmental Panel on Climate Change Fifth Report* (pp. 1–16). Cambridge: European Climate Foundation, the sustainable fisheries partnership, university of Cambridge's judge business school and institute for sustainable leadership. Available online at: https://www.intechopen.com/books/climate-change-and-variability/climate-change-impacts-on-fisheries-and-aquaculture (accessed on 3 August 2024).
78. IPCC. (2013). *Summary for Policymakers, The Physical Science Basis. Contribution of Working Group I to the Fifth Assessment Report of the Intergovernmental Panel on Climate Change*. Cambridge; New York, NY: Cambridge University Press.
79. IPCC. (2014). *Climate Change 2014: Synthesis Report. Contribution of Working Groups I, II and III to the Fifth Assessment Report on the Intergovernmental Panel on Climate Change*. Core writing team: R. K. Pachauri & L. A. Meyer. Geneva: Intergovernmental Panel on Climate Change. Available online at: http://www.ipcc.ch/pdf/assessment-report/ar5/syr/SYR_AR5_FINAL_full_wcover.pdf (accessed on 3 August 2024).
80. IPCC. (2018). *Global Warming of 1.5°C. An IPCC Special Report on the Impacts of Global Warming of 1.5°C Above Pre-Industrial Levels and Related Global Greenhouse Gas Emission Pathways, in the Context*

of Strengthening the Global Response to the Threat of Climate Change, Sustainable Development, and Efforts to Eradicate Poverty (Eds. V. Masson-Delmotte, P. Zhai, H. O. Pörtner, D. Roberts, J. Skea, P. R. Shukla, A. Pirani, W. Moufouma-Okia, C. Péan, R. Pidcock, S. Connors, J. B. R. Matthews, Y. Chen, X. Zhou, M. I. Gomis, E. Lonnoy, T. Maycock, M. Tignor, & T. Waterfield).

81. IPCC. (2019). *Special Report on Climate Change, Desertification, Land Degradation, Sustainable Land Management, Food Security, And Greenhouse Gas Fluxes in Terrestrial Ecosystems, Summary for Policymakers Approved Draft*. Geneva: IPCC.
82. Jahan, A., Nipa, T. T., Islam, S. M. M., Uddin, M. H., Islam, M. S., & Shahjanan, M. (2019). Striped catfish (*Pangasianodon hypophthalmus*) could be suitable for coastal aquaculture. *Journal of Applied Ichthyology, 35*(5), 994–1003. https://doi.org/10.1111/jai.13918.
83. Johnston, P., Everard, M., Santillo, D., & Robert, K. H. (2007). Reclaiming the definition of sustainability. *Environmental Science and Pollution Research, 14*(1), 60–66. https://doi.org/10.1065/espr2007.01.375.
84. Kandu, P. (2017). Papua New Guinea. Impacts of climate variations on local fisheries and aquaculture resources in PNG. In E. J. Ramos (Ed.), *Ecological Risk Assessment of Impacts of Climate Change on Fisheries and Aquaculture Resources* (pp. 45–49). Peru: APEC Ocean and Fisheries Working Group.
85. Kettle, N. P., Dow, K., Tuler, S., Webler, T., Whitehead, J., & Miller, K. M. (2014). Integrating scientific and local knowledge to inform risk-based management approaches for climate adaptation. *Climate Risk Management, 17–31*. https://doi.org/10.1016/j.crm.2014.07.001.
86. Khaine, I., & Woo, S. Y. (2015). An overview of the interrelationship between climate change and forests. *Forest Science and Technology, 11*(1), 11–18. https://doi.org/10.1080/21580103.2014.932718.
87. Khatri-Chhetri, A., Regmi, P. P., Chanana, N., & Aggarwal, P. K. (2019). Potential of climate-smart agriculture in reducing women farmers' drudgery in high climatic risk areas. *Climatic Change, 158*(1), 29–42. https://doi.org/10.1007/s10584-018-2350-8.
88. Kibria, G., Haroon, Y. A. K., & Dayanthi, N. (2017). Climate change impacts on tropical and temperate fisheries, aquaculture, and seafood security and implications: A review. *Livestock Research for Rural Development, 29*(1), 22.
89. King, A. D., & Harrington, L. J. (2018). The inequality of climate change

from 1.5 to 2°C of global warming. *Geophysical Research Letters, 45*(10), 5030–5033. https://doi.org/10.1029/2018GL078430.

90. Klinger, D. H., Levin, S. A., & Watson, J. R. (2017). The growth of finfish in global open-ocean aquaculture under climate change. *Proceedings of the Royal Society B: Biological Sciences, 284*(1862), 20170834. https://doi.org/10.1098/rspb.2017.0834.
91. Kobayashi, M., Msangi, S., Batka, M., Vannuccini, S., Dey, M. M., & Anderson, J. L. (2015). Fish to 2030: The role and opportunity for aquaculture. *Aquaculture Economics & Management, 19*(3), 282–300. https://doi.org/10.1080/13657305.2015.994240.
92. Koblinsky, C. J., Hildebrand, P., LeVine, D., Pellerano, F., Chao, Y., Wilson, W., et al. (2003). Sea surface salinity from space. *Radio Science, 38*(3), 7. https://doi.org/10.1029/2001RS002584.
93. Kroeker, K. J., Gaylord, B., Hill, T. M., Hosfelt, J. D., Miller, S. H., & Sanford, E. (2014). The role of temperature in determining species' vulnerability to ocean acidification: A case study using *Mytilus galloprovincialis*. *PLoS ONE, 9*(7), e100353. https://doi.org/10.1371/journal.pone.0100353.
94. Lafferty, K. D. (2009). The ecology of climate change and infectious diseases. *Ecology, 90*(4), 888–900. https://doi.org/10.1890/08-0079.1.
95. Laffoley, D. (2013). Introduction to the special issue: The global state of the ocean; interactions between stresses, impacts and some potential solutions. Synthesis papers from the International Programme on the State of the Ocean 2011 and 2012 workshops. *Marine Pollution Bulletin, 74*(2), 491–494. https://doi.org/10.1016/j.marpolbul.2013.06.057.
96. Leal Filho, W. (2011). *Experiences of Climate Change Adaptation in Africa*. Berlin; Heidelberg: Springer.
97. Lemasson, A. J., Hall-Spencer, J. M., Fletcher, S., Provstgaard-Morys, S., & Knights, A. M. (2018). Indications of future performance of native and non-native adult oysters under acidification and warming. *Marine Environmental Research, 142*, 178–189. https://doi.org/10.1016/j.marenvres.2018.10.003.
98. Lesperance, A. (2017). Indigenous traditional knowledge remains marginalized in Canada's healthcare system. *Canada's Healthcare System*. Available online at: https://www.abs-canada.org/featured/indigenous-traditional-knowledge-remains-marginalized-in-canadas-healthcare-system (accessed on 3 August 2024).
99. Lim-Camacho, L., Hobday, A. J., Bustamante, R. H., Farmery, A., Fleming, A., Frusher, S., et al. (2014). Facing the wave of change: Stakeholder

perspectives on climate adaptation for Australian seafood supply chains. *Regional Environmental Change, 15*(4), 595–606. https://doi.org/10.1007/s10113-014-0670-4.

100. Magrin, G. O., Marengo, J. A., Boulanger, J. P., Buckeridge, M. S., Castellanos, E., Poveda, G., et al. (2014). Central and South America. In V. R. Barros, C. B. Field, D. J. Dokken, M. D. Mastrandrea, K. J. Mach, T. E. Bilir, et al. (Eds.), *Climate Change 2014: Impacts, Adaptation, and Vulnerability. Part B: Regional Aspects. Contribution of Working Group II to the Fifth Assessment Report of the Intergovernmental Panel on Climate Change* (pp. 1499–1566). Cambridge; New York, NY: Cambridge University Press.
101. Makondo, C. C., & Thomas, D. S. G. (2018). Climate change adaptation: Linking indigenous knowledge with western science for effective adaptation. *Environmental Science & Policy, 88*, 83–91. https://doi.org/10.1016/j.envsci.2018.06.014.
102. Malcorps, W., Kok, B., van't Land, M., Fritz, M., van Doren, D., Servin, K., et al. (2019). The sustainability conundrum of fishmeal substitution by plant ingredients in shrimp feeds. *Sustainability, 11*(4), 1212. https://doi.org/10.3390/su11041212.
103. Mangan, S., Urbina, M. A., Findlay, H. S., Wilson, R. W., & Lewis, C. (2017). Fluctuating seawater pH/pCO2 regimes are more energetically expensive than static pH/pCO2 levels in the mussel *Mytilus edulis*. *Proceedings of the Royal Society B: Biological Sciences, 284*(1868), 20171642. https://doi.org/10.1098/rspb.2017.1642.
104. Marcogliese, D. J. (2008). The impact of climate change on the parasites and infectious diseases of aquatic animals. *Revue Scientifique et Technique de l'Office International des Epizooties, 27*(2), 467–484. https://doi.org/10.20506/rst.27.2.1820.
105. Marshall, N. A., Marshall, P. A., Tamelander, J., Obura, D., Malleret-King, D., & Cinner, J. E. (2009). A framework for social adaptation to climate change: Sustaining tropical coastal communities [SIC] and industries. Available online at: https://portals.iucn.org/library/efiles/documents/2010-022.pdf (accessed on 3 August 2024).
106. Maulu, S., & Musuka, C. G. (2018). Assessing the abundance and distribution of *Tilapia* species in Lake Kariba. *International Journal of Fisheries and Aquaculture Sciences, 8*, 1–11.
107. Maulu, S., Munganga, B. P., Hasimuna, O. J., Haambiya, L. H., & Seemani, B. (2019). A review of the science and technology developments in Zambia's aquaculture industry. *Journal of Aquaculture Research &*

Development, 10(567). https://doi.org/10.4172/2155-9546.1000.567.

108. Merino, G., Barange, M., Blanchard, J. L., Harle, J., Holmes, R., Allen, I., et al. (2012). Can marine fisheries and aquaculture meet fish demand from a growing human population in a changing climate? *Global Environmental Change, 22*(4), 795–806. https://doi.org/10.1016/j.gloenvcha.2012.03.003.
109. Mohanty, B. P., Mohanty, S., Sahoo, J. K., & Sharma, A. P. (2010). Climate change: Impacts on fisheries and aquaculture. In *Climate Change and Variability* (Chapter 7). Zagreb; Rijeka: InTech. https://doi.org/10.5772/9805.
110. Moore, S., Trainer, V., Mantua, N., Parker, M., Laws, E., & Backer, L. (2008). Impacts of climate variability and future climate change on harmful algal blooms and human health. *Environmental Health, 7*(S2), S4. https://doi.org/10.1186/1476-069X-7-S2-S4.
111. Morash, A. J., & Alter, K. (2015). Effects of environmental and farm stress on abalone physiology: Perspectives for abalone aquaculture in the face of global climate change. *Reviews in Aquaculture, 8*(4), 342–368. https://doi.org/10.1111/raq.12097.
112. Morris, J. P., Backeljau, T., & Chapelle, G. (2019). Shells from aquaculture: A valuable biomaterial, not a nuisance waste product. *Reviews in Aquaculture, 11*(1), 42–57. https://doi.org/10.1111/raq.12225.
113. Munday, P. L., Dixson, D. L., Donelson, J. M., Jones, G. P., Pratchett, M. S., Devitsina, G. V., et al. (2009). Ocean acidification impairs olfactory discrimination and homing ability of a marine fish. *Proceedings of the National Academy of Sciences U.S.A., 106*(6), 1848–1852. https://doi.org/10.1073/pnas.0809996106.
114. Munday, P. L., Hernaman, V., Dixson, D. L., & Thorrold, S. R. (2011). Effect of ocean acidification on otolith development in larvae of a tropical marine fish. *Biogeosciences, 8*(5), 1631–1641. https://doi.org/10.5194/bg-8-1631-2011.
115. Myers, S. S., Smith, M. R., Guth, S., Golden, C. D., Vaitla, B., Mueller, N. D., et al. (2017). Climate change and global food systems: Potential impacts on food security and undernutrition. *Annual Review of Public Health, 38*, 259–277. https://doi.org/10.1146/annurev-publhealth-031816-044356.
116. Ngoan, L. D. (2018). Effects of climate change in aquaculture: Case study in Thua Thien Hue Province, Vietnam. *Biomedical Journal of Scientific & Technical Research, 10*(3), 2018. https://doi.org/10.26717/BJSTR.2018.10.001892.
117. Nguyen, L. A., Pham, T. B., Bosma, R., Verreth, J., Leemans, R., De Silva,

S., et al. (2018). Impact of climate change on the technical efficiency of striped catfish, *Pangasianodon hypophthalmus. Aquaculture Economics & Management, 22*(4), 570–581. https://doi.org/10.1111/jwas.12488.

118. Olesen, I., Groen, A. F., & Gjerde, B. (2000). Definition of animal breeding goals for sustainable production systems. *Journal of Animal Science, 78*(3), 570–582. https://doi.org/10.2527/2000.783570x.
119. Osch, S. V., Hynes, S., Freeman, S., & O'Higgins, T. (2019). Estimating the public's preferences for sustainable aquaculture: A country comparison. *Sustainability, 11*(3), 569. https://doi.org/10.3390/su11030569.
120. Palmer, T., & Stevens, B. (2019). The scientific challenge of understanding and estimating climate change. *Proceedings of the National Academy of Sciences U.S.A., 116*(49), 24390–24395. https://doi.org/10.1073/pnas.1906691116.
121. Parker, R. W. R., Blanchard, J. L., Gardner, C., Green, B. S., Hartmann, K., Tyedmers, P. H., et al. (2018). Fuel use and greenhouse gas emissions of world fisheries. *Nature Climate Change, 8*(4), 333–337. https://doi.org/10.1038/s41558-018-0117-x.
122. Paukert, C. P., Lynch, A. J., & Whitney, J. E. (2016). Effects of climate change on North American inland fishes: Introduction to the special issue. *Fisheries, 41*(7), 329–330. https://doi.org/10.1080/03632415.2016.1187011.
123. Pickering, T. D., Ponia, B., Hair, C. A., Southgate, P. C., Poloczanska, E. S., Della Patrona, L., et al. (2011). Vulnerability of aquaculture in the tropical Pacific to climate change. In J. D. Bell, J. E. Johnson, & A. J. Hobday (Eds.), *Vulnerability of Tropical Pacific Fisheries and Aquaculture to Climate Change* (pp. 647–731). Noumea: Secretariat of the Pacific Community.
124. Pongthanapanich, T., Nguyen, A. T., & Xinhua, Y. (2016). Insurance for fishery and aquaculture adaptation to climate change: Experiences from China and Vietnam. Rome: FAO.
125. Pörtner, H. O., Langenbuch, M., & Reipschläger, A. (2004). Biological impact of elevated ocean CO2 concentrations: Lessons from animal physiology and earth history. *Journal of Oceanography, 60*(4), 705–718. https://doi.org/10.1007/s10872-004-5763-0.
126. Preto, B. L., Henares, M. N. P., Kimpara, J. M., & Valenti, W. C. (2015). Greenhouse gas emissions from aquaculture: Aquaculture might have potential for carbon fi□ation. *Global Aquaculture Advocate*. Available online at: https://www.aquaculturealliance.org/advocate/greenhouse-emissions-from-aquaculture/ (accessed on 3 August 2024).

127. Richards, R. G., Davidson, A. T., Meynecke, J., Beattie, K., Hernaman, V., Lynam, T., et al. (2015). Effects and mitigations of ocean acidification on wild and aquaculture scallop and prawn fisheries in Queensland, Australia. *Fisheries Research, 161*, 42–56. https://doi.org/10.1016/j.fishres.2014.06.013.

128. Riphah, S. U. (2015). Global warming: Causes, effects and solutions. *Durreesamin Journal, 1*(1), 1–7.

129. Robinson, R. A., Learmonth, J. A., Hutson, A. M., Macleod, C. D., Sparks, T. H., Leech, D. I., et al. (2005). Climate change and migratory species. *British Trust for Ornithology*. Thetford: The Nunnery.

130. Rodrick, G. B. (2008). Effects of temperature, salinity and pesticides on oyster hemocyte activity. *Florida Water Resources Journal, 4*, 14.

131. Rodrigues, L. C., Van Den Bergh, J. J. M., Massa, F., Theodorou, J. A., Ziveri, P., & Gazeau, A. F. (2015). Sensitivity of Mediterranean bivalve mollusc aquaculture to climate change, ocean acidification, and other environmental pressures: Findings from a producer survey. *Journal of Shellfish Research, 34*(4), 1161–1176. https://doi.org/10.2983/035.034.0341.

132. Rosenzweig, C., Elliott, J., Deryng, D., Ruane, A. C., Muller, C., Arneth, A., et al. (2014). Assessing agricultural risks of climate change in the 21st century in a global gridded crop model intercomparison. *Proceedings of the National Academy of Sciences U.S.A., 111*(9), 3268–3273. https://doi.org/10.1073/pnas.1222463110.

133. Rutkayova, J., Vácha, F., Maršálek, M., Beneš, K., Civišová, H., Horká, P., et al. (2017). Fish stock losses due to e□treme floods–findings from pond-based aquaculture in the Czech Republic. *Journal of Flood Risk Management, 11*(3), 351–359. https://doi.org/10.1111/jfr3.12332.

134. Saba, V. S., Griffies, S. M., Anderson, W. G., Winton, M., Ale□ander, M. A., Delworth, T. L., et al. (2016). Enhanced warming of the Northwest Atlantic Ocean under climate change. *Journal of Geophysical Research: Oceans, 121*(1), 118–132. https://doi.org/10.1002/2015JC011346.

135. Sae-Lim, P., Kause, A., Mulder, H. A., & Olesen, I. (2017). Breeding and genetics symposium: Climate change and selective breeding in aquaculture. *Journal of Animal Science, 95*(4), 1801–1812. https://doi.org/10.2527/jas2016.1066.

136. Schjolden, A. (2004). Towards assessing socioeconomic impacts of climate change in Norway: Sensitivity in the primary sectors: Fisheries, agriculture, agriculture and forestry. *CICERO Report 2004:03*. Oslo: Center for International Climate and Environmental Research.

137. Seggel, A., De Young, C., & Soto, D. (2016). Climate change implications for fisheries and aquaculture: Summary of the findings of the Intergovernmental Panel on Climate Change Fifth Assessment Report. *FAO Fisheries and Aquaculture Circular No. 1122*. Rome: FAO.

138. Shelton, C. (2014). Climate change adaptation in fisheries and aquaculture–Compilation of initial examples. *FAO Fisheries and Aquaculture Circular No. 1088*. Rome: FAO.

139. Silbiger, N. J., & Sorte, C. J. B. (2018). Biophysical feedbacks mediate carbonate chemistry in coastal ecosystems across spatiotemporal gradients. *Scientific Reports, 8*, 796. https://doi.org/10.1038/s41598-017-18736-6.

140. Sondak, C. F. A., Ang, P. O., Beardall, J., Bellgrove, A., Boo, S. M., Gerung, G. S., et al. (2016). Carbon dioxide mitigation potential of seaweed aquaculture beds (SABs). *Journal of Applied Phycology, 29*(4), 2363–2373. https://doi.org/10.1007/s10811-016-1022-1.

141. Stévant, P., Rebours, C., & Chapman, A. (2017). Seaweed aquaculture in Norway: Recent industrial developments and future perspectives. *Aquaculture International, 25*(6), 1373–1390. https://doi.org/10.1007/s10499-017-0120-7.

142. Stewart, H. A., Aboagye, D. I., Ramee, S. W., & Allen, P. J. (2019). Effects of acute thermal stress on acid-base regulation, haematology, ion-osmoregulation and aerobic metabolism in Channel Catfish (*Ictalurus punctatus*). *Aquaculture Research, 50*(8), 2133–2141. https://doi.org/10.1111/are.14093.

143. Sult, V. T., Aldon, M. E. T., Tendencia, I. T., Ortiz, A. M. J., Alayon, S. B., & Ledesma, A. S. (2005). *Regional Technical Consultation on the Aquaculture of P. vannamei and Other Exotic Shrimps in Southeast Asia*. Manila: SEAFDEC/AQD.

144. Thomsen, J., Stapp, L. S., Haynert, K., Schade, H., Danelli, M., Lanning, G., et al. (2017). Naturally acidified habitat selects for ocean acidification tolerant mussels. *Science Advances, 3*(6), e1602411. https://doi.org/10.1126/sciadv.1602411.

145. Tiller, R., De Kok, J. L., Vermeiren, K., Richards, R., Ardelan, M. V., & Bailey, J. (2016). Stakeholder perceptions of links between environmental changes to their socio-ecological system and their adaptive capacity in the region of Troms, Norway. *Frontiers in Marine Science, 3*, 267. https://doi.org/10.3389/fmars.2016.00267.

146. Tisdell, C. A., & Leung, P. S. (1999). Overview of environmental and sustainability issues in aquaculture. *Aquaculture Economics & Management, 3*(1), 1–5. https://doi.org/10.1080/13657309909380228.

147. Toussaint, M., Gyalog, G., Hough, C., & Ytteborg, E. (2018). *The Effects of Climate Change Upon Aquaculture*. Oslo: Climefish.

148. Trainer, V. L., Moore, S. K., Hallegraeff, G., Kudela, R. M., Clement, A., Mardones, J. I., et al. (2019). Pelagic harmful algal blooms and climate change: Lessons from nature's experiments with extremes. *Harmful Algae, 91*, 101591. https://doi.org/10.1016/j.hal.2019.03.009.

149. Troell, M., Eide, A., Isaksen, J., Hermansen, Ø., & Crépin, A. S. (2017). Seafood from a changing Arctic. *Ambio, 46*(S1), S368–S386. https://doi.org/10.1007/s13280-017-0954-2.

150. Troell, M., Naylor, R. L., Metian, M., Beveridge, M., Tyedmers, P. H., Folke, C., et al. (2014). Does aquaculture add resilience to the global food system? *Proceedings of the National Academy of Sciences U.S.A., 111*(37), 13257–13263. https://doi.org/10.1073/pnas.1404067111.

151. Valenti, W. C., Kimpara, J. M., & Preto, B. L. (2011). Measuring aquaculture sustainability. *World Aquaculture, 42*(3), 26–30.

152. Valenti, W. C., Kimpara, J. M., Preto, B. L., & Moraes-Valenti, P. (2018). Indicators of sustainability to assess aquaculture systems. *Ecological Indicators, 88*, 402–413. https://doi.org/10.1016/j.ecolind.2017.12.068.

153. VGREEN. (2012). *Life Cycle Assessment of Fish Feeds: Case Study in Bangladesh*. WorldFish/USAID "Feed the Future-Aquaculture Bangladesh and CSISA Projects. Centre of Excellence on Environment Strategy for GREEN Business (VGREEN), Bangkok: Kasetsart University.

154. Wasmund, N., Nausch, G., & Matthaus, W. (1998). Phytoplankton spring blooms in the Southern North Sea, spatio-temporal development and long-term trends. *Journal of Plankton Research, 20*(6), 1099–1117. https://doi.org/10.1093/plankt/20.6.1099.

155. Weatherdon, L. V., Magnan, A. K., Rogers, A. D., Sumaila, U. R., & Cheung, W. W. L. (2016). Observed and projected impacts of climate change on marine fisheries, aquaculture, coastal tourism, and human health: An update. *Frontiers in Marine Science, 3*, 48. https://doi.org/10.3389/fmars.2016.00048.

156. Wittmann, A. C., & Pörtner, H. O. (2013). Sensitivities of extant animal ta□a to ocean acidification. *Nature Climate Change, 3*(11), 995–1001. https://doi.org/10.1038/nclimate1982.

157. Yazdi, S. K., & Shakouri, D. (2010). The effects of climate change on aquaculture. *International Journal of Environmental Science and Development, 1*(4), 378–382. https://doi.org/10.7763/IJESD.2010.V1.73.

158. Ytrestoyl, T., Aas, T. S., & Åsgård, T. (2015). Utilization of feed resources

in production of Atlantic salmon (*Salmo salar*) in Norway. *Aquaculture, 448*, 365–374. https://doi.org/10.1016/j.aquaculture.2015.06.023.

159. Zhang, P. F., Zhao, T., Zhou, L., Han, G. D., Shen, Y. W., & Ke, C. H. (2019). Thermal tolerance traits of the undulated surf clam *Paphia undulata* based on heart rate and physiological energetics. *Aquaculture, 498*, 343–350. https://doi.org/10.1016/j.aquaculture.2018.08.037.
160. Zolnikov, T. R. (2018). Climate change. In T. R. Zolnikov (Ed.), *Autoethnographies on the Environment and Human Health* (pp. 32–37). Cham: Springer. https://doi.org/10.1007/978-3-319-69026-1.
161. Zolnikov, T. R. (2019). *Global Adaptation and Resilience to Climate Change*. Palgrave Studies in Climate Resilient Societies. Bern. https://doi.org/10.1007/978-3-030-01213-7.

Chapter 5

Climate Smart Aquaculture

Contents

INTRODUCTION

Aquaculture is experiencing substantial growth in its production, making it the most rapidly expanding sector in global food production. Nevertheless, the long-term viability of the industry is in jeopardy as a result of the anticipated consequences of climate change, which are not only a future concern but also a current actuality. This chapter focuses on the potential impacts of climate change on aquaculture production and the resulting implications for the sector's sustainability. Discussions have taken place regarding different aspects of a changing climate, including increasing temperatures, rising sea levels, diseases and harmful algal blooms, alterations in rainfall patterns, uncertainty surrounding external input supplies, changes in sea surface salinity, and severe climatic events. In addition, various strategies for adaptation have been proposed, along with identified areas of insufficient understanding that necessitate further research. In general, climate change is expected to have both negative and positive effects on the sustainability of aquaculture production. However, the negative effects are more significant than the positive ones. The only way to sustain the sector's production is by adapting to the anticipated short-term changes while implementing long-term mitigation measures. However, the success of adaptation will be contingent upon the adaptive capacity of producers in various global regions.

5.1. CONCEPTUAL FRAMEWORK OF CLIMATE SMART AQUACULTURE

Aquaculture is a method used to attain sustainability in the production of aquatic products. Due to the ongoing excessive extraction of fish from natural habitats, the capture fisheries industry is considered the sole means of satisfying the increasing global demand for aquatic products (AskarySary et al., 2012; FAO, 2020). FAO (2020) reports that aquaculture's share of global fish production has steadily increased, reaching 82.1 million tons (46%) out of the estimated 179 million tons of global production. Moreover, the proportion of aquaculture production in relation to the overall global fish production is projected to increase from the current 46% to 53% by the year 2030, according to the FAO (2020). Nevertheless, the primary issue of immediate importance is whether the industry is experiencing sustainable and rapid growth that can adequately meet the anticipated future demand, which is further intensified by the rapidly expanding human population and the impact of climate change. Climate change is currently recognized as a hazard to worldwide food production and a significant danger to the level and amount of production (Beach and Viator, 2008; Hamdan et al., 2015; Myers et al., 2017). The anticipated impacts of climate change are posing a growing threat to food security, specifically in terms of the availability of dietary protein (Kandu, 2017).

Climate change pertains to long-term alterations in the statistical patterns of weather, spanning from decades to millions of years (Yazdi and Shakouri, 2010; IPCC, 2014). These variations can manifest in the average weather or the distribution of weather events around an average. They can be localized to a specific region or occur globally (Yazdi and Shakouri, 2010). Humans are widely acknowledged as the primary cause of climate change due to their reliance on fossil fuels (such as coal, oil, and gas) for energy production. This has been supported by various studies (Doubleday et al., 2013; Environmental Protection Agency, 2016; Gao et al., 2016; Barange et al., 2018; IPCC, 2019, Palmer and Stevens, 2019). Additionally, human activities such as deforestation and forest degradation (Khaine and Woo, 2015; Riphah, 2015) also contribute to the emission of greenhouse gases (GHGs) into the atmosphere. The progressive buildup of greenhouse gases (GHGs) such as carbon dioxide (CO_2), methane (CH_4), nitrous o□ides (N_2O), and fluorinated gases in the atmosphere has been associated with these human activities. Climate change impacts have already been documented in several important global economic sectors and services (Troell et al., 2017; IPCC, 2018; Cook and Zolnikov, 2019; FAO, 2020). Recent literature in aquaculture suggests that there are noticeable changes in climate affecting water resources, including rising temperatures, changing precipitation patterns, and more frequent extreme events. Some of these changes are already evident, while others are still emerging (Fleming et al., 2014; Blanchard et al., 2017; Troell et al., 2017; Zolnikov, 2019). The impact of climate change on the sustainability of aquaculture has become a topic of great interest due to the sector's substantial role in global food security, nutrition, and livelihoods (Blanchard et al., 2017; Dabbadie et al., 2018; FAO, 2020). Although there are some uncertainties, multiple projections indicate that the entire aquaculture value chain is susceptible to the impacts of climate change (Cochrane et al., 2009; Fleming et al., 2014; Bueno and Soto, 2017; Barange et al., 2018; Dabbadie et al., 2018). Most of the existing literature is heavily biased towards demonstrating the impact of climate change on aquaculture production systems, while neglecting to address other stages in the value chain (Fleming et al., 2014). These biases restrict our present comprehension of the diverse degrees to which the aquaculture sector will be impacted. Furthermore, the lack of comprehensive global models that can consider both the immediate and indirect impacts of climate change on aquaculture complicates the process of measuring these effects (Metian, 2017). The effects of this situation will be long-lasting and probably cannot be undone, which will have serious negative impacts on the economy of those involved in the sector. Poorer communities are expected to be particularly affected by these extreme consequences, as indicated by various sources (IPCC, 2013, 2014; Holmyard, 2014; Barange et al., 2018; Dabbadie et al., 2018).

Extensive research and analysis have been conducted on the impact of climate change on aquaculture at both regional and global levels. These studies include the works of De Silva and Soto (2009), Yazdi and Shakouri (2010), Clements and Chopin (2016), Bueno and Soto (2017), Chung et al., (2017), Ellis et al., (2017), Froehlich et al., (2017), Handisyde et al., (2017), Harvey et al., (2017), Klinger et al., (2017), Beveridge et al., (2018), and Dabbadie et al., (2018). Most of these studies have focused on investigating the adverse impacts of climate change on aquaculture, while neglecting the equally important positive effects that are crucial for developing adaptation strategies. An e□tensive e□amination of the adverse and beneficial aspects of climate change will enhance producers' readiness and mitigate the risks to their production. Moreover, the existing literature fails to explore the potential impact of climate change on the sustainability of the aquaculture industry. This information is highly valuable for identifying suitable interventions to address climate change and ensure the long-term sustainability of livelihood sources for communities that rely on aquaculture.

5.2. THE EFFECTS OF CLIMATE CHANGE ON AQUACULTURE AND IMPLICATIONS ON SUSTAINABILITY

The impact of climate change on aquaculture production is anticipated to be both direct and indirect, as stated by Handisyde et al., (2006) and De Silva and Soto (2009). The direct effects of climate change on finfish and shellfish stocks in production systems involve changes to their physical and physiological characteristics. Indirect effects, on the other hand, can occur through alterations to the primary and secondary productivity and structure of ecosystems. These changes can also impact the availability and cost of input supplies, such as fishmeal and fish oil, as well as other goods and services required by fishers and aquaculture producers. (Handisyde et al., 2006; De Silva and Soto, 2009; Freeman, 2017; Adhikari et al., 2018). The subsequent section will provide a comprehensive analysis of the diverse impacts of climate change on aquaculture production and the resulting implications for the sector's sustainability. It is widely agreed that aquaculture production is not isolated, but rather has significant connections with other food production systems (De Silva and Soto, 2009; Troell et al., 2014). In addition, Blanchard et al., (2017) noted that in order to meet the growing demand for aquatic products in a sustainable manner, it is important to acknowledge the interconnectedness between the goals of fisheries, aquaculture, and agriculture systems. Figure 5.1 presents a clear depiction of the impact of greenhouse gas (GHG) emissions on the production of aquaculture, as well as the role of capture fisheries, aquaculture, and agriculture in contributing to climate change.

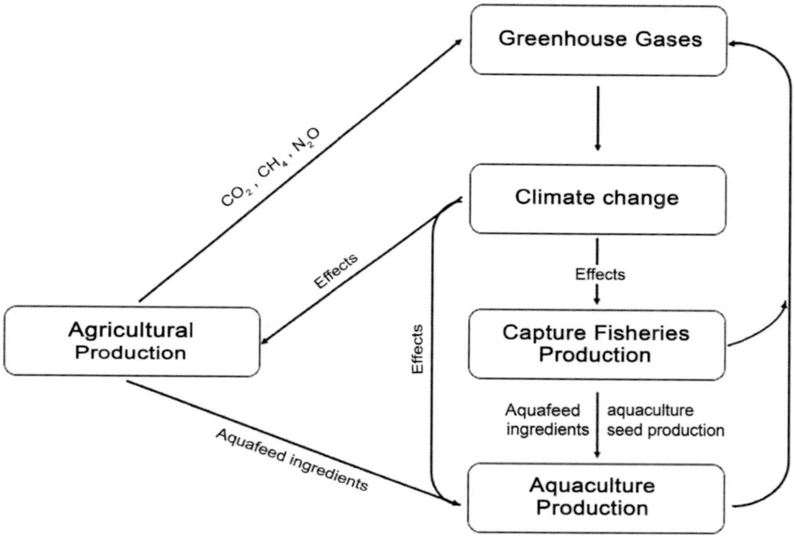

Figure 5.1. *An illustration of the direct and indirect routes by which climate change will impact aquaculture production.*

While aquaculture activities, including power input, transport, and feed production, are recognized as the primary sources of greenhouse gas emissions in the sector (Cochrane et al., 2009; Adhikari et al., 2018; Barange et al., 2018), their overall impact is relatively small compared to other food production sectors, although still significant (Barange et al., 2018). As an illustration, in 2010, aquaculture was responsible for emitting approximately 385 million tons of CO2, which accounted for around 7% of the total greenhouse gas emissions from the agricultural sector (Barange et al., 2018). According to recent estimates by the Intergovernmental Panel on Climate Change (IPCC) in 2019, agriculture, forestry, and other land uses were responsible for approximately 13% of carbon dioxide (CO2) emissions, 44% of methane (CH4) emissions, and 82% of nitrous oxide (N2O) emissions from human activities during the period of 2007-2016. These emissions accounted for 23% of the total human-caused greenhouse gas (GHG) emissions. Animal production, specifically livestock farming, is responsible for approximately 45% of the total net contribution made by agriculture. This sector is the largest contributor of CH4,

CO2, and N2O emissions, according to studies by Preto et al., (2015), De Silva and Soto (2009), and Zolnikov (2018). However, aquaculture primarily emits CO2 as a result of the regular respiration of aquatic animals (De Silva and Soto, 2009). Nevertheless, there remains a deficiency in current understanding regarding the routes and impact of aquaculture production on global greenhouse gas emissions, necessitating additional research. In 2011, the fisheries sector, specifically CO2 emissions, accounted for appro□imately 4% of the total greenhouse gas emissions from food production. Between 1990 and 2011, there was a significant increase of 28% in emissions from the global fisheries industry, primarily driven by e□panded fishing activities (Daw et al., 2009; Parker et al., 2018).

In recent times, aquaculture has experienced notable technological advancements that allow the industry to increase its current production in order to meet the growing demand for aquatic products (Kobayashi et al., 2015; Morris et al., 2019; FAO, 2020). Climate change is emerging as a significant challenge for the long-term viability of food production systems, including aquaculture (Lim-Camacho et al., 2014; IPCC, 2018; FAO, 2020). IPCC (2018) states that sustainable development, which involves achieving a balance between environmental protection, economic prosperity, and social well-being, is closely interconnected with the impacts and actions related to climate change. This implies that without addressing the effects of climate change, it is not possible to achieve sustainability in aquaculture production. Sustainability, as defined by Johnston et al., (2007), encompasses the effective management of financial, technological, institutional, natural, and social resources to ensure a continuous provision of human needs, both in the present and for future generations (Valenti et al., 2018). Hence, in order for aquaculture to be considered sustainable, it must be able to persist over multiple generations (Valenti et al., 2011). Valenti et al., (2018) propose that the evaluation of aquaculture systems' sustainability can be done by considering three key indicators: environmental, economic, and social indicators. Environmental sustainability indicators encompass the effective utilization of natural resources, prevention of pollution, and conservation of biodiversity (Valenti et al., 2018). Economic sustainability pertains to the effective utilization of financial resources, economic viability, resilience, and the ability to offset negative external costs, as well as the generation of funds for reinvestment. On the other hand, social sustainability concerns the ability of aquaculture to deliver advantages to communities, including food security, employment, fair distribution of income and opportunities, and the inclusion of vulnerable populations (Valenti et al., 2011, 2018). Nevertheless, the existing body of literature on aquaculture sustainability has primarily concentrated on environmental sustainability, while giving considerably less consideration to the other two equally significant aspects (Tisdell and Leung, 1999; Ahmed et al., 2019; Engle, 2019). Future research should rigorously examine the potential impact of climate-induced changes on the social and economic dimensions of aquaculture production sustainability. A study conducted in Ghana revealed that climate-induced changes have the potential to substantially reduce the economic worth of aquaculture products and exacerbate poverty, particularly in rural areas (Asiedu et al., 2019). The majority of information found in the literature consists of forecasts derived from scientific models.

Multiple studies have indicated that the impact of climate change on aquaculture can vary depending on factors such as geographical location, economic conditions, climate zones, production systems, and the species being cultured (Merino et al., 2012; Bell et al., 2013; Cheung et al., 2013; Sae-Lim et al., 2017; Adhikari et al., 2018; Barange et al., 2018; IPCC, 2018; Zolnikov, 2019). Barange et al., (2018) projected that producers in developing nations and poorer economies would e□perience more significant impacts compared to those in developed nations. According to Handisyde et al., (2017), the impact of climate change on aquaculture producers is anticipated to vary based on the specific environments in which they operate, such as freshwater, brackish, or marine settings. Additional research has demonstrated that climate change risks will have a greater impact on small-scale farmers compared to large-scale producers. This is primarily due to the increased costs associated with managing farms and the absence of support systems to mitigate the effects of climate change. These findings have been supported by studies conducted by Schjolden (2004), Sult et al., (2005), Hamdan et al., (2012), and King and Harrington (2018). Moreover, it is crucial to acknowledge that the impacts of climate change will not solely impact aquaculture production systems, but also the entire value chain (Fleming et al., 2014; Barange et al., 2018). Therefore, climate change can be seen as an unintentional threat that increases susceptibility to socio-economic development and places significant strain on food demand and supply, as well as the livelihoods of farmers.

The anticipated factors of a changing climate that pose a threat to the production and sustainability of the aquaculture sector include increasing temperature, ocean acidification, diseases and harmful algal blooms, alterations in rainfall and precipitation patterns, rising sea levels, uncertainty regarding external input supplies, changes in sea surface salinity, and severe climatic events. (Handisyde et al., 2006; Brander, 2007; Ficke et al., 2007; Barange et al., 2018). The impact of these elements on aquaculture production varies, as the sector is influenced by factors such as time, space, and size, which provide a certain level of fle□ibility. Additionally, these elements affect fish populations at different stages of their life cycle. In addition, the e□isting scientific understanding of the impact of individual consequences is diverse and often restricted to the cumulative effects, which poses practical challenges for adaptation planning in the aquaculture industry (Seggel et al., 2016).

5.2.1. Rising Temperature

The temperature is a crucial factor that significantly influences the growth and development of aquatic organisms (Ngoan, 2018). Due to their poikilothermic nature, fish may exhibit heightened sensitivity to temperature fluctuations caused by climate change (Sae-Lim et al., 2017; Adhikari et al., 2018). It is expected that the average global temperature will increase by 1.5°C this century. As a result, there is a high probability of increased deaths among various fish species, particularly those that thrive in cold-water environments such as the Atlantic halibut, Salmon, and Cod. Additionally, intertidal shellfish are also likely to experience higher mortality rates due to thermal stress. These findings are supported by studies conducted by Hamdan et al., (2012) and Gubbins et al., (2013). Consequently, extended exposure to high or low temperatures can have negative effects on aquaculture productivity, resulting in reduced output. Chronic stress can impact the neuroendocrine and osmoregulatory systems, causing changes in the performance of the heart and lungs, the ability to perform aerobic activities, and the immune responses of economically important species. This has been shown in studies conducted by Brodie et al., (2014), Gazeau et al., (2014), Paukert et al., (2016), Stévant et al., (2017), Stewart et al., (2019), and Zhang et al., (2019). Moreover, the metabolism, physiology, feeding behavior, and growth performance of the majority of finfish and shellfish species are expected to be influenced (Marcogliese, 2008; Akegbejo-Samsons, 2009; Lemasson et al., 2018). In addition, increasing ocean temperatures and resulting ocean acidification gradually reduce the ability of the ocean to absorb carbon, leading to changes in the flow and characteristics of water systems, as well as the occurrence of red tides (Cochrane et al., 2009). The aforementioned effects can result in escalated operational expenses and diminished efficiency, posing a risk to the economic and social viability of aquaculture production. Thermal stratification in deep water bodies can impact environmental sustainability by causing temperature variations. This can affect the distribution and abundance of nutrients in the water. In the event of upwelling, aquaculture producers operating in open waters will experience significant economic losses (Seggel et al., 2016). Nevertheless, there is a scarcity of data on the physiological reaction of the most economically significant species to increasing temperatures. This data is primarily focused on a small number of species and is predominantly centered around adult stages, leaving the initial developmental stages, such as embryos, larvae, and fingerlings, poorly understood. Conversely, periods of higher temperatures (within the limits that species can tolerate) can extend the duration of growing seasons, particularly in temperate areas, and enhance the growth of species that thrive in warmer water, such as the Giant tiger prawn, Tilapia, Oysters, and Mussels (Pickering et al., 2011; Troell et al., 2017; Guyondet et al., 2018; Collins et al., 2020). Investors who operate hatcheries in protected areas may also profit from market opportunities that arise from the decrease in desired specimens in the wild caused by the deterioration of coral reefs (Bell et al., 2010). Warmer periods in cold regions, like the Arctic, are expected to promote the growth of aquaculture production (Chan et al., 2019). In addition, periods of higher temperatures can create favorable conditions for the cultivation of new species and promote advancements in the genetic enhancement of aquatic organisms (Gubbins et al., 2013; Bueno and Soto, 2017). These opportunities will promote social sustainability by increasing production outputs and employment opportunities, as well as economic sustainability by boosting profits and reducing management costs in these areas. However, attaining this goal will necessitate advancements in molecular biology and the implementation of practical techniques for genetic enhancement in aquaculture. However, it is important to note that this may pose a risk to environmental sustainability if hybridization occurs with species in natural bodies of water.

5.2.2. Ocean Acidification

Ocean acidification is the long-term decrease in pH levels of ocean water caused by the absorption of atmospheric CO2 over a period of several decades (Richards et al., 2015; Bahri et al., 2018). According to Seggel et al., (2016), the oceans are believed to contain approximately 50 times more CO2 than the atmosphere. According to the IPCC (2018), the expected rise in CO2 absorption by oceans due to a global warming of 1.5°C or higher will negatively impact the growth, development, calcification, survival, and abundance of various aquatic species. The heightened buildup of carbon dioxide (CO2) in water can lead to a rise in water acidity levels, resulting in a decrease in pH (Rodrigues et al., 2015; Clements and Chopin, 2016). This poses a threat to the environmental sustainability of aquaculture production systems as it deteriorates water quality, leading to reduced productivity. In addition, the increase in ocean acidity decreases the availability of carbonate, which is necessary for the formation of coral skeletons (calcification) in shell-forming organisms like shrimps, mussels, oysters, and corals. This poses a potential threat to marine aquaculture production. Increased predation rates on juvenile wild spat oysters, caused by poor coral skeleton formation, may lead to a decline in oyster production. This decline is attributed to lower collection rates, as observed in a study by Blanchard et al., (2017). As a result, this is likely to compel major producers to depend on hatcheries for oyster spat production, leading to an increase in production costs. This could result in decreased production outputs and profits, which could have adverse effects on the social and economic sustainability of aquaculture production in these regions. The increasing acidity levels in seawater have the potential to greatly impact the physiology and metabolism of aquatic species by interfering with the mechanisms responsible for transporting substances between cells (Pörtner et al., 2004). The global capture of the ability of species to respond to changes in ocean acidity is challenging (Froehlich et al., 2018). This ability depends on the species' adaptive capacity (Thomsen et al., 2017), the rate at which changes occur (Cooley et al., 2012; Mangan et al., 2017), and complex biophysical feedbacks (Silbiger and Sorte, 2018). The production of macroalgal (seaweed) may be influenced by ocean acidification. However, the specific impact will vary depending on how different species absorb inorganic carbon, as stated by Chung et al., (2017). Calcifying species in areas where CO2 is not the primary cause of acidification are expected to experience the greatest impact, according to Cornwall et al., (2012) and Clements and Chopin (2016). While seaweed is recognized as a significant absorber of atmospheric CO2, there have been limited investigations into the impact of climate change on seaweed production. This is likely due to the fact that seaweed production is restricted to specific geographical areas. According to Gubbins et al., (2013), accurately predicting future changes in ocean carbonate chemistry is challenging, and it is difficult to explain the effects of these changes due to the complexities of monitoring long-term biological responses to small observed changes in experimental conditions. Moreover, most of the existing studies on the impacts of ocean acidification on aquaculture species have been carried out in conjunction with the effects of increasing temperatures (Rodrigues et al., 2015).

Ocean acidification can have a beneficial effect by leading to a significant economic advantage for hatchery owners, as large-scale aquaculture producers may become increasingly reliant on hatcheries for spat production. Moreover, the establishment of these hatcheries is likely to generate increased job opportunities for local communities due to the substantial labor needed to meet the rising demand. This, in turn, will contribute to the social and economic sustainability of aquaculture production. The impact of ocean acidification on finfish is not well understood. However, the presence of calcified otoliths in finfish, especially marine species, makes them vulnerable to ocean acidification. This can affect their growth, development, RNA viability, and cause damage to tissues, as well as impair respiration.

5.2.3. Diseases and Harmful Algal Blooms

The impact of changing temperature regimes on diseases in aquaculture, including bacterial, parasitic, viral, and fungal diseases, is expected to be variable and difficult to predict (Collins et al., 2020). It is indisputable that cultured species become more vulnerable to diseases when they are exposed to thermal stress conditions. Additionally, warmer conditions can lead to the emergence of unfamiliar diseases (Collins et al., 2020). The susceptibility of finfish and shellfish to pathogens is a significant factor in the occurrence of diseases and is expected to be influenced by both direct and indirect thermal stressors (Chiaramonte et al., 2016). Consequently, the frequency of warm water disease outbreaks is expected to increase, along with the potential for uncovering previously unknown outbreaks, due to climate change (Sae-Lim et al., 2017). According to Marcogliese (2008), an increase in temperature is expected to speed up the reproduction rate, strength, lifespan, and spread of disease-causing organisms among various types of fish and shellfish. In addition, the rising temperature pressures can facilitate the occurrence of epizootic diseases in aquaculture and lead to significant economic difficulties. The outbreak of epizootic diseases is currently a significant constraint on the success of aquaculture production systems in numerous countries worldwide (Maulu et al., 2019). Additionally, it is anticipated that the presence of warm water pathogens, such as Sea lice, will continue to pose difficulties in salmon farming. Furthermore, as temperatures rise, the infections in cold temperate environments are likely to worsen, necessitating increased treatments and subsequently higher expenses (Collins et al., 2020). In summary, the rising prevalence of diseases in aquaculture production systems will result in decreased profits, thereby impacting the social and economic sustainability of aquaculture production. On the other hand, cold water diseases like vibriosis and winter ulcer, which harm Atlantic salmon, could potentially disappear over time because of the worsening conditions (Sae-Lim et al., 2017) that may benefit the growth of this fish species.

The outbreak of certain harmful algal blooms has been linked to changes in climatic conditions, as demonstrated by various studies (Wasmund et al., 1998; Edwards and Richardson, 2004; Edwards et al., 2006; Moore et al., 2008; De Silva and Soto, 2009; Lafferty, 2009; Trainer et al., 2019). Algal blooms pose a significant risk to the long-term viability of aquaculture production in terms of environmental sustainability. For instance, taxonomic groups such as flagellates and dinoflagellates, along with other harmful species, have been found to harbor potentially toxic or nuisance species. These species can cause stress or mortality in finfish and shellfish. This has been reported by Hinder et al., (2012), Gubbins et al., (2013), and Basti et al., (2019). As a result, this could have adverse effects on the social and economic aspects of maintaining the sustainability of aquaculture. A recent report documented an unparalleled decline in fish populations in Chilean aquaculture. This decline was attributed to the proliferation of Pseudochattonella cf. verruculosa and Alexandrium catenella species, whose outbreak was linked to climate-induced alterations in water column stratification (Trainer et al., 2019). In addition, various studies have documented the presence of abnormalities, such as inflammation, shrinkage, and tissue death, in multiple organs of bivalve mollusks caused by harmful algal blooms (Haberkorn et al., 2010; Basti et al., 2011; Hégaret et al., 2012). While the specific effects of climate change on toxic substances in aquaculture are not well-documented, Farrell et al., (2015) found that changes in temperature can impact the metabolism of commonly occurring harmful algae.

5.2.4. Changes in Rainfall (Precipitation) Pattern

Fluctuations in precipitation will have contrasting impacts on aquaculture production and sustainability: heightened rainfall leading to flooding, and periods of reduced or absent rainfall resulting in drought. Based on the IPCC (2018) report, the probability of increased risks from droughts is higher at 2°C of global warming compared to 1.5°C in a specific region. On the other hand, it is challenging to accurately predict patterns of flooding events. Higher precipitation levels, especially if they occur as more intense events, will amplify the risks associated with agricultural production in low-lying regions (Bell et al., 2010). The risks associated with fish ponds include the loss of fish during floods, the invasion of unwanted species into the ponds, and damage to the ponds caused by infilling and the erosion of walls (Rutkayova et al., 2017). The introduction of invasive fish species and deterioration of water quality are the main ways in which the mixing of pond water and fish with those in the wild can harm the environmental sustainability of aquaculture production. In addition, the depletion of fish populations in ponds poses a significant risk to the social and economic aspects of maintaining sustainable aquaculture. This is because it reduces the financial benefits for producers and can lead to poverty within communities. Rutkayova et al., (2017) found that the percentage of fish lost during periods of severe flooding varies depending on the species and age of the individual fish. The authors also noted that the percentage of losses is expected to decrease as the age of the fish increases. It is important to acknowledge that increased rainfall can expand the suitable areas for aquaculture ponds that depend on rainwater in low-lying tropical regions. This can have a positive impact on the social and economic sustainability of these regions (Bell et al., 2013). According to Collins et al., (2020), increased rainfall can impact the productivity of macroalgae, such as kelp, by introducing different amounts of nutrients into nearshore environments. The fluctuation in nutrient loading caused by varying precipitation patterns may promote the growth of invasive short-lived algal species instead of the longer-lived kelp species that are typically preferred for cultivation (Gubbins et al., 2013).

Drought events can result in water scarcity, including shortages and degradation of water quality, which can have adverse impacts on aquaculture production (Hambal et al., 1994). The anticipated water scarcity resulting from climate change will result in heightened disputes over water resources among various user groups, including aquaculture, agriculture, domestic, and industrial sectors (Handisyde et al., 2006; Barange et al., 2018). This will have an impact on all aspects of aquaculture sustainability. Nevertheless, it is imperative to conduct additional research to e□amine the specific reactions of various fish species and their different life stages, particularly those that hold economic significance, to alterations in precipitation patterns.

5.2.5. Sea Level Rise

The IPCC (2018) projects that sea levels will be approximately 0.1 meters lower under a 1.5°C global warming scenario compared to a 2°C scenario by the year 2100. Nevertheless, it is anticipated that this increase will persist beyond the year 2100, with the extent and speed of the rise expected to be influenced by the future trajectories of greenhouse gas emissions (IPCC, 2018). The increase in sea level has the potential to devastate various coastal ecosystems, including mangroves and salt marshes, which are vital for preserving wild fish populations and providing seeds for aquaculture production (Kibria et al., 2017). This will have a detrimental impact on aquaculture breeding programs and the long-term economic viability of the sector. The intrusion of saline water caused by higher sea levels is expected to have a significant impact on aquaculture production facilities, including ponds, cages, tanks, and pens, especially in lowland regions (Kibria et al., 2017). The process of salinization of groundwater is considered detrimental to aquaculture, freshwater fisheries, and agricultural production (Handisyde et al., 2006; Kibria et al., 2017). As a result, salinization makes aquaculture not suitable for production from an environmental perspective, which in turn increases production costs and reduces economic gains. Sea level rise is expected to cause alterations in the types of species present, the abundance and distribution of organisms, the productivity of ecosystems, and changes in the timing of natural events, which may pose a risk to both inland and marine aquaculture production (Doney et al., 2012). In addition, aquaculture activities in coastal regions yield social and environmental advantages that can be directly or indirectly impacted by the rising sea levels, consequently influencing the productivity and long-term viability of the sector. Sea-level rise may have a positive impact by expanding the suitable areas for brackish water culture of valuable species, such as shrimp and mud crab (Handisyde et al., 2006; Kibria et al., 2017). This could enhance the sustainability of aquaculture production by offering fresh possibilities, especially for aquaculture operations located in coastal regions.

5.2.6. Uncertainty of External Input Supplies

Aquaculture production heavily relies on external inputs from agriculture and capture fisheries, indicating a significant interdependence between these systems. According to Cochrane et al., (2009), aquaculture is a supplementary activity to capture fisheries. While it shares similarities with agriculture in its methods, it also has significant connections with capture fisheries. Although agriculture currently serves as the primary source of ingredients for energy needs in aquatic animal feeds and is expected to continue being the main provider of protein sources in the future, capture fisheries currently play a crucial role in supplying protein sources, as well as wild seed and broodstock for aquaculture (Hardy, 2010). In recent times, there has been a shift towards using cereal and soy production in aquaculture feed manufacturing, as a result of the decreasing fish catches from capture fisheries (Ytrestoyl et al., 2015). Nevertheless, the susceptibility of agricultural production to the impacts of climate change poses a significant risk, thereby jeopardizing the availability of resources necessary to sustain aquaculture production (Khatri-Chhetri et al., 2019). The effects of climate change on capture fisheries, specifically changes in fish species abundance and distribution, will greatly impact the availability of fishmeal and fish oil. This will also affect the sustainability of fish breeding programs due to the increasing scarcity of wild seeds. These findings are supported by studies conducted by Bradley et al., (2015), Blanchard et al., (2017), and Bueno and Soto (2017). The inadequate governance of fisheries and escalating costs of fishmeal pose a substantial risk to the long-term viability of aquaculture production (Black and Hughes, 2017). In addition, the anticipated effects of climate change on fishery resources are expected to hasten the inadequate management of capture fisheries (Black and Hughes, 2017; Barange, 2019).

In general, climate change is predicted to have a negative effect on agriculture and capture fisheries. This will result in a decrease in the availability and an increase in the cost of inputs, such as fish seed and feed ingredients, that are necessary for aquaculture production. As a result, the expenses associated with aquaculture production are anticipated to increase, posing challenges for small-scale producers to sustain their operations in the industry. Conversely, the increasing costs of fishmeal and fish oil are e□pected to e□pedite advancements in scientific research aimed at finding substitute protein and oil sources to replace traditional ingredients in aquafeeds (Hardy, 2010). In recent times, there has been significant interest in utilizing plant proteins, particularly oilseeds derived from agriculture, as protein sources in aquaculture. Currently, there is a growing demand to find alternative ingredients to replace fishmeal and fish oils in feed formulations for the aquaculture sector in order to ensure its long-term sustainability.

5.2.7. Changes in Sea Surface Salinity

Salinity is a variable parameter that indicates the amount of freshwater input from precipitation, ice melting, river runoff, water loss through evaporation, and the mixing and circulation of ocean surface water with underground water (Koblinsky et al., 2003; Cochrane et al., 2009). Changes in sea salinity can arise from elevated evaporation caused by higher temperatures and alterations in ocean circulation, or can be directly influenced by climate change (Cooper, 1988; Robinson et al., 2005). These fluctuations can impact the movement and layering of ocean currents, which in turn affects the ocean's ability to retain heat, as well as the circulation of carbon and nutrients (Seggel et al., 2016). Given that climate change is anticipated to induce these fluctuations, the ecological viability of aquaculture will be impacted. Aquatic organisms typically have narrow ranges of salinity tolerance, and any deviations from their optimal salinity levels can result in significant mortality rates and reduced production yields (Jahan et al., 2019). Excessive salinity levels in stripped catfish have been found to decrease their survival, growth, and red blood cell count, indicating an impact on the fish's immune system (Jahan et al., 2019). Fluctuations in sea salinity are anticipated to have a detrimental impact on the economic profitability of certain aquaculture species, potentially compromising the social and economic aspects of sustainable aquaculture production. Nevertheless, there is a significant correlation between increased salinity and the aquaculture production systems in coastal areas downstream (Nguyen et al., 2018). Ahmed (2013) documented adverse impacts on the production performance of freshwater prawns when exposed to elevated salinity levels. Baker et al., (2005) observed a higher death rate in young clams, whereas Rodrick (2008) found that oysters were more prone to bacterial invasion when exposed to lower water salinity. In addition, Rodrick (2008) noted that changes in salinity can impact the immune system of oysters, specifically the capacity of hemocytes (blood cells) to defend against bacterial intrusion from outside sources. While clams are known to have a broader range of salinity tolerance compared to other marine mollusks, Baker et al., (2013) observed that they are unable to withstand prolonged exposure to either high or low salinity levels. Predictions have been made that abalone mortalities in farms, especially those located in coastal areas, will increase due to the intrusion of saline water from marine environments as a result of climate change (Morash and Alter, 2015). Typically, fluctuations in water salinity will result in higher death rates for multiple species, which can have negative impacts on the economic and social viability of the sector due to increased species losses and higher management expenses.

Nevertheless, most of the existing knowledge regarding the impact of climate-related changes on salinity in aquaculture has been inclined towards documenting the consequences of elevated salinity. Research is necessary to investigate the impact of suboptimal salinity levels on both finfish and shellfish. In addition, our understanding of how various commercially valuable species respond to changes in salinity caused by climate change is limited. The provided information is particularly valuable for implementing adjustments in aquaculture, as variations in salinity can promote the growth of species that are tolerant to such changes (Jahan et al., 2019).

5.2.8. Severe Climatic Events

Severe climatic events, such as cyclones, waves, and storms, are anticipated to have an impact on the growth of aquaculture, particularly in the production of marine ornamental products and in coastal regions (Toussaint et al., 2018). For instance, coral and giant clam farmers in tropical villages may encounter the possibility of higher losses due to bleaching, while those in sub-tropical regions are expected to face more significant risks, such as the loss of production equipment and stock caused by rougher sea conditions associated with stronger cyclones (Hamdan et al., 2012). Storm surges, waves, and coastal erosion pose significant risks to aquaculture production and other coastal activities, making them the most dangerous threats (Hamdan et al., 2012). Severe storms will cause significant financial losses for farmers due to the damage inflicted on their farms, resulting in increased costs for recovering from the damage (Canadian Institute for Climate Studies, 2000). For example, cage aquaculture is expected to encounter increased vulnerabilities due to the impacts of cyclones and waves. The projected increase in storminess during specific seasons and in specific regions may heighten the likelihood of aquatic organism escapes caused by equipment failure. This could necessitate the relocation of aquaculture sites or modifications in production practices, potentially posing significant challenges to the social and economic sustainability of aquaculture in these areas (Gubbins et al., 2013). Barange et al., (2018) found that severe climatic events have become more frequent in various regions recently, accounting for at least 80% of all climate-related disasters. Nevertheless, it is projected that these occurrences will become more frequent in Africa, specifically in East and Southern Africa (IPCC, 2019).

However, intense weather events, such as storms, are expected to have a significant impact on the mi□ing of water layers and nutrients that were previously limited to specific layers due to thermal stratification (Seggel et al., 2016). This could potentially enhance the environmental sustainability of aquaculture production. In addition, storms can play a critical role in reducing water temperatures and mitigating potential risks to both cultivated and wild organisms (Seggel et al., 2016).

5.3. MITIGATION AND ADAPTATION OPTIONS

The projected 1.5°C global warming will lead to an increase in climate change-related risks to health, food security, livelihood, water supply, human security, and economic development. These risks are expected to further increase at 2°C (IPCC, 2018). To address these risks, both the industry and communities must reduce and adjust to the changing climate by capitalizing on new opportunities arising from modified resources (Lorenzen et al., 2017). Implementing mitigation and adaptation strategies can enhance the preparedness of farming communities, ecosystems, and populations in addressing climate change with maximum effectiveness and efficiency (Zolnikov, 2019). Mitigation aims to decrease or reverse the rate of climate change (Leal Filho, 2011; ACT, 2018). This primarily entails the reduction of greenhouse gas (GHG) emissions, with a particular emphasis on carbon dioxide (CO2) emissions. CO2 emissions are responsible for over 60% of the human-induced increase in GHGs (Mohanty et al., 2010; IPCC, 2014; Environmental Protection Agency, 2016). The IPCC (2018) states that a mix of new and existing technologies and practices, such as electrification, hydrogen, sustainable bio-based feedstocks, product substitution, and carbon capture utilization and storage, can be used to decrease CO2 emissions. Producers and other stakeholders in aquaculture can have a substantial impact on reducing the effects of climate change by implementing necessary changes in their production practices to minimize the release of greenhouse gases (GHGs). More precisely, this encompasses the utilization of eco-friendly methods and technology, such as solar power, appropriate feeding techniques, and sustainable wastewater treatment, to reduce air and water contamination (VGREEN, 2012; Barange et al., 2018). The production of feed in aquaculture is widely recognized as the primary source of greenhouse gas emissions in the sector (VGREEN, 2012). For instance, sinking feeds have been found to be more ecologically sustainable when compared to floating feeds, as reported by Hardy (2010) and VGREEN (2012). Nevertheless, mitigation is a strategy for addressing climate change that requires a considerable amount of time before its advantages can be fully recognized (Elum et al., 2017; ACT, 2018). In addition, successful mitigation necessitates collaborative efforts on a worldwide level as the majority of greenhouse gases (GHGs) accumulate gradually and disperse globally, leading to a universal impact (IPCC, 2014). Hence, it is recommended that mitigation and adaptation strategies be implemented in conjunction to achieve improved and efficient outcomes (ACT, 2018; IPCC, 2019).

Adaptation primarily aims to develop the ability to effectively cope with the outcomes and take advantage of new possibilities in a manner that is environmentally friendly and morally sound (Bueno and Soto, 2017). Strategic foresight encompasses the process of anticipating and incorporating anticipated changes into both immediate decision-making and long-term planning (Yazdi and Shakouri, 2010). Hence, it can manifest in diverse forms, such as technical modifications, alterations in the conduct of resource users/producers, or modifications in the governance system (Lorenzen et al., 2017).

Furthermore, according to FAO (2018), there are three specific areas that successful adaptation interventions can focus on: institutions and management, livelihood adaptation, and resilience and risk reduction. Nevertheless, it is highly probable that a synthesis of these domains will produce superior outcomes in contrast to a narrow focus on a single domain, due to the robust interconnection and mutually reinforcing effects between them.

At present, adaptation is regarded as the most feasible choice available to producers for dealing with the impacts of climate change on aquaculture (Yazdi and Shakouri, 2010). In addition, the IPCC (2014) observed that the level of adaptation to the effects of climate change is increasing in various regions, both in the public and private sectors. This recognition and inclusion of adaptation measures in developmental plans by several governments is also on the rise. Moreover, it is advisable to prioritize adaptation options that can also be utilized for mitigating greenhouse gas (GHG) emissions.

This is because these options have the potential to create synergies that can result in cost reductions across various sectors (IPCC, 2018). Adaptation is a complex process that cannot be applied universally. It depends on the interplay between vulnerability, exposure, and the socio-economic and sustainable development of different sectors.

This has been highlighted by various sources such as the IPCC (2014, 2018), Clements and Chopin (2016), and Zolnikov (2019). As per the IPCC (2018), the ability of producers in a specific nation or region to adapt to a changing climate will determine the success of their adaptation efforts. For instance, researchers have forecasted that producers in developing countries will experience more severe consequences as a result of their limited ability to adapt to changes. This is supported by multiple studies (Handisyde et al., 2017; Barange et al., 2018; IPCC, 2018; King and Harrington, 2018; Zolnikov, 2019) and is illustrated in Figure 5.2. Adaptation options are more likely to be effective and sustainable when chosen based on a country's specific circumstances and supportive conditions (Bradley et al., 2015; Blanchard et al., 2017; IPCC, 2018).

Expanding the range of livelihoods can be crucial for effectively adapting to climate change. This approach provides producers with multiple options to sustain their livelihoods and develop resilience against climate change impacts (Bell et al., 2013; Duarte et al., 2017; Zolnikov, 2019). It entails the integration of aquaculture production systems with other sectors, such as agricultural systems, either as integrated or separate systems.

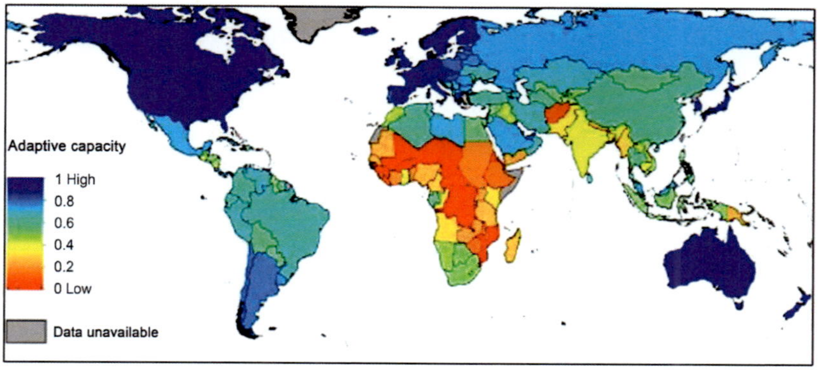

Figure 5.2. *World map showing the adaptive capacity of the nations to climate change effects on aquaculture.*

E□panding the range of income-generating activities is highly beneficial, particularly in regions or countries where there is a projected decrease in fish production and an anticipated increase in agricultural production (Blanchard et al., 2017). Nevertheless, in order to achieve successful diversification, it is imperative that government policies offer incentives to ensure the efficient allocation of resources, fairness, and environmental preservation (Troell et al., 2014). Furthermore, the enhancement of livelihood diversification among aquaculture producers necessitates the integration of indigenous knowledge with government interventions (Leal Filho, 2011). The primary determinant of the ability of farmers, households, and communities to adapt is their access to, control over, and productive utilization of various assets such as natural resources, human capital, social networks, physical infrastructure, and financial resources. These assets collectively form the livelihood capital (Bueno and Soto, 2017).

Aquaculture producers could potentially gain advantages by transitioning to aquaculture species, techniques, or locations that are less susceptible or more resistant to the impacts of a changing environment and limited resources (Lim-Camacho et al., 2014). The Integrated Multi-Trophic Aquaculture (IMTA) system is a more environmentally friendly, sustainable, and economically rewarding approach compared to monoculture. It is also more resilient to changing climate conditions. This is because IMTA combines finfish farming with other species, as demonstrated by Osch et al., (2019). Aquaculture producers can utilize modified resources, such as land, through the implementation of innovative farming systems. Moreover, transitioning to or cultivating aquaculture species that exhibit greater resistance to the effects of climate change is considered one of the most encouraging strategies for adaptation (De Young et al., 2012). Sae-Lim et al., (2017) observed that in a changing climate, the selection of species based on feed efficiency and the establishment of a breeding goal that minimizes greenhouse gas emissions will be crucial. Nevertheless, it is necessary to take into account the environmental and social advantages of communities rather than solely prioritizing immediate economic gains (Olesen et al., 2000). Additionally, it is crucial to acknowledge that there will be expenses, such as research and development, linked to the creation of novel aquaculture species (Harvey et al., 2017).

One additional method for aquaculture producers to adapt to climate change is to guarantee a consistent supply of fish from capture fisheries. This can be achieved by encouraging modifications in fishers' behavior, implementing changes in governance, and employing effective management plans and strategies. For instance, implementing modifications to the administration of the transferable allowance assigned to individual fishermen (Frusher et al., 2013). Furthermore, it is necessary to integrate climate variability and change into the modeling of aquaculture activities in order to mitigate the effects of climate change on livelihoods dependent on fisheries. Based on the FAO (2020) report, the worldwide production from capture fisheries has remained at a constant level or experienced a decrease in certain years throughout the previous few decades. The escalating global population and subsequent surge in demand for aquatic products are regarded as the main catalysts for the declining production from capture fisheries (Maulu and Musuka, 2018). At present, aquaculture production is heavily dependent on resources obtained from capture fisheries, including fishmeal, fish oils, broodstocks, and wild seeds, which are essential for production (Laffoley, 2013; FAO, 2018; Malcorps et al., 2019). Hence, proficient administration of resources derived from fisheries can contribute to a sustainable provision of inputs for aquaculture.

The integration and application of local and indigenous knowledge in adaptation strategies have demonstrated their efficacy in addressing climate change (Leal Filho, 2011; IPCC, 2014; Magrin et al., 2014; Makondo and Thomas, 2018). Scientific understanding offers a broad understanding of climate change and its possible impacts, while indigenous knowledge can provide specific information about the physical environment, infrastructure systems, livelihood status, behavior, governance organization, and other important factors for managing community resources (Kettle et al., 2014). For instance, employing conventional techniques and indicators to forecast alterations in weather patterns such as temperature, rainfall, humidity, etc., can assist producers in anticipating anticipated changes and developing resilience (Zolnikov, 2019). The extent to which this incorporation has been utilized in aquaculture production remains largely unknown. Nevertheless, in various industries, including agriculture, farming communities have effectively implemented it. An illustrative instance can be found in Kenya, where the utilization of indigenous knowledge enabled certain farming communities to proactively anticipate the impacts of climate change and mitigate their vulnerability to food insecurity (Leal Filho, 2011). In addition, the combination of indigenous knowledge and scientific knowledge has the potential to greatly enhance other adaptation strategies, even though it has been neglected in the past (Kettle et al., 2014; Belfer et al., 2017; Lesperance, 2017; Makondo and Thomas, 2018).

One potential area for adaptation is enhancing adaptive capacity in aquaculture, particularly for small-scale producers, by implementing insurance schemes. Most climate change predictions indicate that small-scale producers will be the most affected due to poor adaptive capacity (IPCC, 2014; Bueno and Soto, 2017; Barange et al., 2018). Therefore, implementing an insurance program could assist them in developing the ability to recover quickly from difficulties. Generally, insurance for aquaculture producers globally is still relatively new despite gaining considerable attention (Pongthanapanich et al., 2016). In Vietnam, a pilot project conducted between 2011 and 2013 showed that farmers had a very poor understanding of the scheme which led to poor results (Pongthanapanich et al., 2016), suggesting that awareness among other measures may lead to better results. Insurance schemes may especially be useful for producers with poor adaptive capacity for building resilience against the effects of climate change (Barange et al., 2018). However, the viability of the aquaculture insurance business may depend on how efficient and lower-risk aquaculture becomes, apart from climate change risks (Barange et al., 2018). Being a relatively new subject in aquaculture, there is a need for investigations to e□pound its benefits and implications on the economic conditions of the farmers, especially among the most vulnerable communities.

5.3.1. Challenges to Successful Adaptation and Proposed Strategies

While adaptation is widely regarded as the most feasible short-term response to climate change, its execution may encounter various obstacles. Failure to recognize the fact that climate change will impact producers in varying ways depending on factors such as their specific region, environment, knowledge of climate change, population, community dynamics, economic conditions, and current industry (Marshall et al., 2009). If these factors are not considered, adaptation strategies are likely to provide minimal advantages to the producers and stakeholders in the aquaculture sector (Zolnikov, 2019). In order to tackle this issue, it is crucial for governments, particularly in countries that are more susceptible to climate change, to ensure that their policy-making process regarding climate change interventions acknowledges this reality and caters to the specific requirements of producers in their respective circumstances. Hence, it is imperative to allocate additional efforts and resources towards the producer groups that are most susceptible to vulnerability. According to Lorenzen et al., (2017), successful adaptation is more probable when producers are informed about the factors and ways in which it affects them, monitor a wide range of impact indicators, and make adaptive decisions. Various models have been employed in recent years to forecast the effects of climate change on food production systems, including aquaculture (Rosenzweig et al., 2014; Saba et al., 2016; Tiller et al., 2016; Blanchard et al., 2017; Galbraith et al., 2017). Incorrect projections can hinder the aquaculture producers' ability to successfully adapt to the effects of climate change. The ability of producers to adjust to climate change will probably rely on precise future forecasts for various aquaculture sector systems. Exaggerated and prejudiced assessment of the potential hazards linked to climate change could deceive policymakers and aquaculture producers in implementing essential preventive measures. Maladaptation leads to a higher likelihood of future vulnerability and/or exposure of the target community, region, or sector to the effects of climate change (IPCC, 2014). In order to mitigate biases in climate models, Palmer and Stevens (2019) emphasize the importance of minimizing the reliance on sub grid parameterizations. If this is not possible, it is crucial to explicitly document the uncertainty associated with these parameterizations. The authors propose that future models should prioritize sustainability and have a global reach, promoting the efficient use of computing and technology. Another obstacle to achieving successful adaptation is the ability of producers to be adequately prepared. For instance, certain regions in the world, particularly those in western countries, possess superior resources in terms of governance, incomes, technology, human capital, and social networks compared to sub-Saharan African countries (Zolnikov, 2019). In addition, various forecasts indicate that aquaculture producers in developing nations will experience the greatest impact, primarily due to their limited ability to adapt (Barange et al., 2018; Zolnikov, 2019). Hence, the IPCC (2018) strongly advocates for international collaboration that facilitates improved financial and technological access, as well as enhanced local capabilities, in order to enable developing nations and the most vulnerable regions to take effective action. It is increasingly evident that the entire aquaculture value chain is susceptible to the impacts of climate change. However, a significant portion of the existing literature primarily emphasizes the production system while neglecting other stages, such as the trade and marketing of aquatic products. The limited scope of scientific studies hinders our comprehension of the potential impact on the aquaculture industry and, consequently, the available options for adaptation. Hence, it is imperative for forthcoming research and models to adopt a more comprehensive approach that encompasses all stages of the aquaculture value chain.

5.4. CONCLUSION

This chapter has emphasized the possible impacts of climate change on aquaculture production and the consequences for the sector's long-term viability. Although the aquaculture sector is seen as the sole answer to the growing global demand for aquatic products, it is facing an escalating threat from the effects of human-induced climate change, which are already happening and will continue to worsen in the future. The impact on aquaculture is anticipated to have both favorable and unfavorable consequences, with the latter outweighing the former. In addition, while climate change poses a global threat to food production, the specific risks to aquaculture are anticipated to vary based on factors such as geographical or climatic zones, national economy, water environment, production systems, scale of production, and the species being cultured by aquaculture producers. In order to maintain productivity and withstand the impacts of climate change, aquaculture producers must promptly adapt to the existing alternatives, while also implementing long-term adjustments to their production practices to minimize the effects. This chapter has discussed significant facets of climate change and aquaculture production. However, it has also identified several limitations that need to be taken into account for future research. Specifically, the chapter focuses solely on the production and input supply stages of the aquaculture value chain, omitting any discussion of other significant stages like trade, processing, and consumption of the products. Furthermore, during the production phase of the value chain under consideration, it remains uncertain how different species of economic significance will adapt to climate change at different stages of their life cycle. Obtaining this information would be valuable for adaptation strategies that may necessitate producers transitioning to species that exhibit greater resilience to climate change. Furthermore, there were limited instances where concrete illustrations of the anticipated impact of climate change on the sustainability of aquaculture production were provided. Additionally, the majority of studies exhibited a partiality towards examining the environmental aspects of sustainability, while disregarding the social and economic dimensions. Given the increasing growth of the aquaculture sector and the growing understanding of climate change, it is necessary to adopt a comprehensive approach to predict the effects of climate change on aquaculture and to effectively deal with these consequences. As a result, the implementation of mitigation and adaptation strategies would yield greater effectiveness. Nevertheless, achieving this goal will necessitate advancements in scientific inquiry, particularly in regions that are more susceptible to adverse impacts due to their limited ability to adapt. Consequently, less affluent economies stand to gain greater advantages from international collaboration.

REFERENCES

1. Adhikari, S., Keshav, C. A., Barlaya, G., Rathod, R., Mandal, R. N., Ikmail, S., et al. (2018). Adaptation and mitigation strategies of climate change impact in freshwater aquaculture in some states of India. *Journal of Fisheries, 12*, 016–021.
2. Ahmed, N. (2013). Linking prawn and shrimp farming towards a green economy in Bangladesh: Confronting climate change. *Ocean & Coastal Management, 75*, 33–42. https://doi.org/10.1016/j.ocecoaman.2013.01.002.
3. Ahmed, N., Thompson, S., & Glaser, M. (2019). Global aquaculture productivity, environmental sustainability, and climate change adaptability. *Environmental Management, 63*(2), 159. https://doi.org/10.1007/s00267-018-1117-3.
4. Akegbejo-Samsons, Y. (2009). Climate change impacts on fisheries production in land-water interface. *IOP Conference Series: Earth and Environmental Science, 6*, 292022. https://doi.org/10.1088/1755-1307/6/29/292022.
5. Asiedu, B., Malcolm, D., & Iddrisu, S. (2019). Assessing the economic impact of climate change in the small-scale aquaculture industry of Ghana, West Africa. *AAS Open Research, 1*, 26. https://doi.org/10.12688/aasopenres.12911.2.
6. AskarySary, A., Velayatzadeh, M., & KarimiSary, V. (2012). Proximate composition of farmed fish, *Oncorhynchus mykiss* and *Cyprinus carpio* from Iran. *Advances in Environmental Biology, 6*(10), 2841–2845.
7. Bahri, T., Barange, M., & Moustahfid, H. (2018). Chapter 1: Climate change and aquatic systems. In M. Barange, T. Bahri, M. C. M. Beveridge, K. L. Cochrane, S. Funge-Smith, & F. Poulaine (Eds.), *Impacts of Climate Change on Fisheries and Aquaculture: Synthesis of Current Knowledge, Adaptation and Mitigation Options* (pp. 1–17). Rome: FAO.
8. Baker, S. M., Baker, P., Heuberger, D., & Sturmer, L. (2005). Short-term effects of rapid salinity reduction in seed clams (*Mercenaria mercenaria*). *Journal of Shellfish Research, 24*(1), 29–33. https://doi.org/10.2983/0730-8000(2005)24[29:SEORSR]2.0.CO;2.
9. Baker, S., Hoover, E., & Sturmer, L. (2013). The role of salinity in hard clam aquaculture. Gainesville, FL: University of Florida.
10. Barange, M. (2019). Avoiding misinterpretation of climate change projections of fish catches: Food for thought. *ICES Journal of Marine Science, 76*(5), 1390–1392. https://doi.org/10.1093/icesjms/fsz061.

11. Barange, M., Bahri, T., Beveridge, M. C. M., Cochrane, A. L., Funge-Smith, S., & Paulain, F. (2018). *Impacts of Climate Change on Fisheries and Aquaculture: Synthesis of Current Knowledge, Adaptation and Mitigation Options*. Rome: FAO.
12. Basti, L., Endo, M., & Segawa, S. (2011). Physiological, pathological, and defense alterations in Manila clams (*Ruditapes philippinarum*) induced by *Heterocapsa circularisquama. Journal of Shellfish Research, 30*(3), 829–844. https://doi.org/10.2983/035.030.0324.
13. Basti, L., Nagai, K., Segawa, S., Tanaka, Y., Suzuki, T., & Nagai, S. (2019). Harmful algal blooms and shellfish aquaculture in changing environment. *Bulletin of the Japan Fisheries Research and Education Agency, No. 49*, 73–79.
14. Beach, R. H., & Viator, C. L. (2008). The economics of aquaculture insurance: An overview of the U.S. pilot insurance program for cultivated clams. *Aquaculture Economics & Management, 12*(1), 25–38. https://doi.org/10.1080/13657300801959613.
15. Belfer, E., Ford, J. D., & Maillet, M. (2017). Representation of Indigenous peoples in climate change reporting. *Climatic Change, 145*(1–2), 57–70. https://doi.org/10.1007/s10584-017-2076-z.
16. Bell, J. D., Ganachaud, A., Gehrke, P. C., Griffiths, S. P., Hobday, A. J., Hoegh-Guldberg, O., et al. (2013). Mi□ed responses of tropical Pacific fisheries and aquaculture to climate change. *Nature Climate Change, 3*(6), 591–599. https://doi.org/10.1038/nclimate1838.
17. Bell, J., Batty, M., Ganachaud, A., Gehrke, P., Hobday, A., Hoegh-Guldberg, O., et al. (2010). Preliminary assessment of the effects of climate change on fisheries and aquaculture in the Pacific. In R. Gillett (Ed.), *Fisheries in the Economies of the Pacific Island Countries and Territories* (pp. 451–469). Manila: Asian Development Bank.
18. Beveridge, M. C. M., Dabbadie, L., Soto, D., Ross, L. G., Bueno, P. B., & Aguilar-Manjarrez, J. (2018). Chapter 22: Climate change and aquaculture: Interactions with fisheries and agriculture. In *Impacts of Climate Change on Fisheries and Aquaculture: Synthesis of Current Knowledge, Adaptation and Mitigation Options* (pp. 461–482). Rome: FAO.
19. Black, K., & Hughes, A. (2017). *Future of the Sea: Trends in Aquaculture, Foresight–Future of the Sea Evidence Review Foresight*. Edinburgh: Government Office for Science.
20. Blanchard, J. L., Watson, R. A., Fulton, E. A., Cottrell, R. S., Nash, K. L., Bryndum-Buchholz, A., et al. (2017). Linked sustainability challenges and

trade-offs among fisheries, aquaculture, and agriculture. *Nature Ecology & Evolution, 1*(9), 1240–1249. https://doi.org/10.1038/s41559-017-0258-8.

21. Bradley, M., Putten, I., & Sheaves, M. (2015). The pace and progress of adaptation: Marine climate change preparedness in Australia's coastal communities. *Marine Policy, 53*, 13–20. https://doi.org/10.1016/j.marpol.2014.11.004.
22. Brander, K. M. (2007). Global fish production and climate change. *Proceedings of the National Academy of Sciences U.S.A., 104*(50), 19709–19714. https://doi.org/10.1073/pnas.0702059104.
23. Brodie, J., Williamson, C. J., Smale, D. A., Kamenos, N. A., Mieszkowska, N., Santos, R., et al. (2014). The future of the northeast Atlantic benthic flora in a high CO_2 world. *Ecology and Evolution, 4*(14), 2787–2798. https://doi.org/10.1002/ece3.1105.
24. Bueno, P. B., & Soto, D. (2017). *Adaptation Strategies of the Aquaculture Sector to the Impacts of Climate Change*. Rome: FAO.
25. Canadian Institute for Climate Studies. (2000). *Aquaculture and Climate Change in Canada: A Discussion Paper*. Ottawa, ON: Canadian Institute for Climate Studies.
26. Chan, F. T., Stanislawczyk, K., Sneekes, A. C., Dvoretsky, A., Gollasch, S., Minchin, D., et al. (2019). Climate change opens new frontiers for marine species in the Arctic: Current trends and future invasion risks. *Global Change Biology, 25*(1), 25–38. https://doi.org/10.1111/gcb.14469.
27. Cheung, W. W. L., Watson, R., & Pauly, D. (2013). Signature of ocean warming in global fisheries catch. *Nature, 497*(7449), 365–369. https://doi.org/10.1038/nature12156.
28. Chiaramonte, L., Munson, D., & Trushenski, J. (2016). Climate change and considerations for fish health and fish health professionals. *Fish Health Section, Fisheries*. Available online at: www.fisheries.org (accessed on 3 August 2024).
29. Chung, I. K., Oak, J. H., Lee, J. A., Shin, J. A., Kim, J. G., & Park, K. S. (2013). Installing kelp forest/seaweed beds for mitigation and adaptation against global warming: Korean project overview. *ICES Journal of Marine Science, 70*(5), 1038–1044. https://doi.org/10.1093/icesjms/fss206.
30. Chung, I. K., Sondak, C. F. A., & Beardall, J. (2017). The future of seaweed aquaculture in a rapidly changing world. *European Journal of Phycology, 52*(4), 495–505. https://doi.org/10.1080/09670262.2017.1359678.
31. Clements, J. S., & Chopin, T. (2016). Ocean acidification and marine

aquaculture in North America: Potential impacts and mitigation strategies. *Reviews in Aquaculture, 9*(4), 326–341. https://doi.org/10.1111/raq.12140.

32. Cochrane, K., De Young, C., Soto, D. T., & Bahri, D. T. (2009). *Climate Change Implications for Fisheries and Aquaculture: Overview of Current Scientific Knowledge*. FAO Fisheries and Aquaculture Technical Paper No. 530. Rome: FAO.
33. Collins, C., Bresnan, E., Brown, L., Falconer, L., Guilder, J., Jones, L., et al. (2020). Impacts of climate change on aquaculture. *MCCIP Science Review 2020*, 482–520. https://doi.org/10.14465/2020.arc21.aqu.
34. Cook, D., & Zolnikov, T. R. (2019). Antarctica. In T. R. Zolnikov (Ed.), *Global Adaptation and Resilience to Climate Change* (pp. 31–49). Palgrave Pivot. https://doi.org/10.1007/978-3-030-01213-7_1.
35. Cooley, S. R., Lucey, N., Kite-Powell, H., & Doney, S. C. (2012). Nutrition and income from molluscs today imply vulnerability to ocean acidification tomorrow. *Fish and Fisheries, 13*(2), 182–215. https://doi.org/10.1111/j.1467-2979.2011.00424.x.
36. Cooper, N. S. (1988). The effect of salinity on tropical ocean models. *Journal of Physical Oceanography, 18*(5), 697–707. https://doi.org/10.1175/1520-0485(1988)018<0697>2.0.CO;2.
37. Cornwall, C. E., Hepburn, C. D., Pritchard, D., Currie, K. I., McGraw, C. M., Hunter, K. A., et al. (2012). Carbon use strategies in macroalgae: Differential responses to lowered pH and implications for ocean acidification. *Journal of Phycology, 48*(1), 137–144. https://doi.org/10.1111/j.1529-8817.2011.01085.x.
38. Dabbadie, L., Aguilar-Manjarrez, J. J., Beveridge, M. C. M., Bueno, P. B., Ross, L. G., & Soto, D. (2018). Chapter 20: Effects of climate change on aquaculture: Drivers, impacts and policies. In *Impacts of Climate Change on Fisheries and Aquaculture: Synthesis of Current Knowledge, Adaptation and Mitigation Options* (pp. 387–408). Rome: FAO.
39. IPCC (2019). *Special Report on Climate Change, Desertification, Land Degradation, Sustainable Land Management, Food Security, and Greenhouse Gas Fluxes in Terrestrial Ecosystems, Summary for Policymakers Approved Draft*. Geneva: IPCC.
40. Jahan, A., Nipa, T. T., Islam, S. M. M., Uddin, M. H., Islam, M. S., & Shahjanan, M. (2019). Striped catfish (*Pangasianodon hypophthalmus*) could be suitable for coastal aquaculture. *Journal of Applied Ichthyology, 35*(6), 994–1003. https://doi.org/10.1111/jai.13918.
41. Johnston, P., Everard, M., Santillo, D., & Robert, K. H. (2007). Reclaiming the definition of sustainability. *Environmental Science & Pollution*

Research, 14(1), 60–66. https://doi.org/10.1065/espr2007.01.375.

42. Kandu, P. (2017). Papua New Guinea: Impacts of climate variations on local fisheries and aquaculture resources in PNG. In E. J. Ramos (Ed.), *Ecological Risk Assessment of Impacts of Climate Change on Fisheries and Aquaculture Resources* (pp. 45–49). Peru: APEC Ocean and Fisheries Working Group.
43. Kettle, N. P., Dow, K., Tuler, S., Webler, T., Whitehead, J., & Miller, K. M. (2014). Integrating scientific and local knowledge to inform risk-based management approaches for climate adaptation. *Climatic Risk Management, 1*, 17–31. https://doi.org/10.1016/j.crm.2014.07.001.
44. Khaine, I., & Woo, S. Y. (2015). An overview of the interrelationship between climate change and forests. *Forest Science and Technology, 11*(1), 11–18. https://doi.org/10.1080/21580103.2014.932718.
45. Khatri-Chhetri, A., Regmi, P. P., Chanana, N., & Aggarwal, P. K. (2019). Potential of climate-smart agriculture in reducing women farmers' drudgery in high climatic risk areas. *Climatic Change, 158*(1), 29–42. https://doi.org/10.1007/s10584-018-2350-8.
46. Kibria, G., Haroon, Y. A. K., & Dayanthi, N. (2017). Climate change impacts on tropical and temperate fisheries, aquaculture, and seafood security and implications–A review. *Livestock Research for Rural Development, 29*(22).
47. Marcogliese, D. J. (2008). The impact of climate change on the parasites and infectious diseases of aquatic animals. *Revue Scientifique et Technique de l'Office International des Epizooties, 27*(2), 467–484. https://doi.org/10.20506/rst.27.2.1820.
48. Marshall, N. A., Marshall, P. A., Tamelander, J., Obura, D., Malleret-King, D., & Cinner, J. E. (2009). A framework for social adaptation to climate change: Sustaining tropical coastal communities and industries. Available online at: https://portals.iucn.org/library/efiles/documents/2010-022.pdf (accessed on 3 August 2024).
49. Maulu, S., & Musuka, C. G. (2018). Assessing the abundance and distribution of Tilapia species in Lake Kariba. *International Journal of Fisheries and Aquaculture Sciences, 8*, 1–11.
50. Maulu, S., Munganga, B. P., Hasimuna, O. J., Haambiya, L. H., & Seemani, B. (2019). A review of the science and technology developments in Zambia's aquaculture industry. *Journal of Aquaculture Research & Development, 10*(567). https://doi.org/10.4172/2155-9546.1000.567.
51. Merino, G., Barange, M., Blanchard, J. L., Harle, J., Holmes, R., Allen, I., et al. (2012). Can marine fisheries and aquaculture meet fish demand

from a growing human population in a changing climate? *Global Environmental Change, 22*(4), 795–806. https://doi.org/10.1016/j.gloenvcha.2012.03.003.

52. Metian, M. (2017). *Aquaculture and Global Changes*. Retrieved from: http://www.ocean-climate.org/wp-content/uploads/2017/03/aquaculture_ScientificNotes_07-14.pdf (accessed on 3 August 2024).
53. Parker, R. W. R., Blanchard, J. L., Gardner, C., Green, B. S., Hartmann, K., Tyedmers, P. H., et al. (2018). Fuel use and greenhouse gas emissions of world fisheries. *Nature Climate Change, 8*(4), 333–337. https://doi.org/10.1038/s41558-018-0117-x.
54. Paukert, C. P., Lynch, A. J., & Whitney, J. E. (2016). Effects of climate change on North American inland fishes: Introduction to the special issue. *Fisheries Magazine, 41*(6), 329–330. https://doi.org/10.1080/03632415.2016.1187011.
55. Pickering, T. D., Ponia, B., Hair, C. A., Southgate, P. C., Poloczanska, E. S., Della Patrona, L., et al. (2011). Vulnerability of aquaculture in the tropical Pacific to climate change. In J. D. Bell, J. E. Johnson, & A. J. Hobday (Eds.), *Vulnerability of Tropical Pacific Fisheries and Aquaculture to Climate Change* (pp. 647–731). Noumea: Secretariat of the Pacific Community.
56. Pongthanapanich, T., Nguyen, A. T., & Xinhua, Y. (2016). *Insurance for Fishery and Aquaculture Adaptation to Climate Change Experiences from China and Vietnam*. Rome: FAO.
57. Pörtner, H. O., Langenbuch, M., & Reipschläger, A. (2004). Biological impact of elevated ocean CO2 concentrations: Lessons from animal physiology and earth history. *Journal of Oceanography, 60*(5), 705–718. https://doi.org/10.1007/s10872-004-5763-0.
58. Richards, R. G., Davidson, A. T., Meynecke, J., Beattie, K., Hernaman, V., Lynam, T., et al. (2015). Effects and mitigations of ocean acidification on wild and aquaculture scallop and prawn fisheries in Queensland, Australia. *Fisheries Research, 161*, 42–56. https://doi.org/10.1016/j.fishres.2014.06.013.
59. Rodrigues, L. C., Van Den Bergh, J. J. M., Massa, F., Theodorou, J. A., Ziveri, P., & Gazeau, A. F. (2015). Sensitivity of Mediterranean bivalve mollusc aquaculture to climate change, ocean acidification, and other environmental pressures: Findings from a producer survey. *Journal of Shellfish Research, 34*(4), 1161–1176. https://doi.org/10.2983/035.034.0341.
60. Rosenzweig, C., Elliott, J., Deryng, D., Ruane, A. C., Muller, C., Arneth, A., et al. (2014). Assessing agricultural risks of climate change in the 21st

century in a global gridded crop model intercomparison. *Proceedings of the National Academy of Sciences of the United States of America, 111*(9), 3268–3273. https://doi.org/10.1073/pnas.1222463110.

61. Rutkayova, J., Vácha, F., Maršálek, M., Beneš, K., Civišová, H., Horká, P., et al. (2017). Fish stock losses due to e□treme floods: Findings from pond-based aquaculture in the Czech Republic. *Journal of Flood Risk Management, 11*(3), 351–359. https://doi.org/10.1111/jfr3.12332.
62. Saba, V. S., Griffies, S. M., Anderson, W. G., Winton, M., Ale□ander, M. A., Delworth, T. L., et al. (2016). Enhanced warming of the Northwest Atlantic Ocean under climate change. *Journal of Geophysical Research: Oceans, 121*(1), 118–132. https://doi.org/10.1002/2015JC011346.
63. Sae-Lim, P., Kause, A., Mulder, H., & Olesen, I. (2017). Breeding and genetics symposium: Climate change and selective breeding in aquaculture. *Journal of Animal Science, 95*(4), 1801–1812. https://doi.org/10.2527/jas2016.1066.
64. Schjolden, A. (2004). *Towards Assessing Socioeconomic Impacts of Climate Change in Norway: Sensitivity in the Primary Sectors: Fisheries, Agriculture, and Forestry*. CICERO Report 2004:03. Oslo: Center for International Climate and Environmental Research.
65. Seggel, A., De Young, C., & Soto, D. (2016). *Climate Change Implications for Fisheries and Aquaculture: Summary of the Findings of the Intergovernmental Panel on Climate Change Fifth Assessment Report*. FAO Fisheries and Aquaculture Circular No. 1122. Rome: FAO.
66. Shelton, C. (2014). *Climate Change Adaptation in Fisheries and Aquaculture–Compilation of Initial Examples*. FAO Fisheries and Aquaculture Circular No. 1088. Rome: FAO.
67. Silbiger, N. J., & Sorte, C. J. B. (2018). Biophysical feedbacks mediate carbonate chemistry in coastal ecosystems across spatiotemporal gradients. *Scientific Reports, 8*(1), 796. https://doi.org/10.1038/s41598-017-18736-6.
68. Sult, V. T., Aldon, M. E. T., Tendencia, I. T., Ortiz, A. M. J., Alayon, S. B., & Ledesma, A. S. (2005). *Regional Technical Consultation on the Aquaculture of P. vannamei and Other Exotic Shrimps in Southeast Asia*. Manila: SEAFDEC/AQD.
69. Thomsen, J., Stapp, L. S., Haynert, K., Schade, H., Danelli, M., Lanning, G., et al. (2017). Naturally acidified habitat selects for ocean acidification tolerant mussels. *Science Advances, 3*(2), e1602411. https://doi.org/10.1126/sciadv.1602411.
70. Tiller, R., De Kok, J. L., Vermeiren, K., Richards, R., Ardelan, M. V., &

Bailey, J. (2016). Stakeholder perceptions of links between environmental changes to their socio-ecological system and their adaptive capacity in the region of Troms, Norway. *Frontiers in Marine Science, 3*, 267. https://doi.org/10.3389/fmars.2016.00267.

71. Tisdell, C. A., & Leung, P. S. (1999). Overview of environmental and sustainability issues in aquaculture. *Aquaculture Economics & Management, 3*(1), 1–5. https://doi.org/10.1080/13657309909380228.
72. Toussaint, M., Gyalog, G., Hough, C., & Ytteborg, E. (2018). *The Effects of Climate Change Upon Aquaculture*. Oslo: Climefish.
73. Trainer, V. L., Moore, S. K., Hallegraeff, G., Kudela, R. M., Clement, A., Mardones, J. I., et al. (2019). Pelagic harmful algal blooms and climate change: Lessons from nature's experiments with extremes. *Harmful Algae, 91*, 101591. https://doi.org/10.1016/j.hal.2019.03.009.
74. Troell, M., Naylor, R. L., Metian, M., Beveridge, M., Tyedmers, P. H., Folke, C., et al. (2014). Does aquaculture add resilience to the global food system? *Proceedings of the National Academy of Sciences of the United States of America, 111*(43), 13257–13263. https://doi.org/10.1073/pnas.1404067111.
75. Valenti, W. C., Kimpara, J. M., & Preto, B. L. (2011). Measuring aquaculture sustainability. *World Aquaculture, 42*, 26–30.
76. Valenti, W. C., Kimpara, J. M., Preto, B. L., & Moraes-Valenti, P. (2018). Indicators of sustainability to assess aquaculture systems. *Ecological Indicators, 88*, 402–413. https://doi.org/10.1016/j.ecolind.2017.12.068.
77. VGREEN. (2012). *Life cycle assessment of fish feeds: Case study in Bangladesh*. WorldFish/USAID "Feed the Future-Aquaculture Bangladesh and CSISA Projects." Centre of E□cellence on Environment Strategy for GREEN Business (VGREEN), Bangkok: Kasetsart University.
78. Wasmund, N., Nausch, G., & Matthaus, W. (1998). Phytoplankton spring blooms in the Southern North Sea, spatio-temporal development and long-term trends. *Journal of Plankton Research, 20*(6), 1099–1117. https://doi.org/10.1093/plankt/20.6.1099.
79. Weatherdon, L. V., Magnan, A. K., Rogers, A. D., Sumaila, U. R., & Cheung, W. W. L. (2016). Observed and projected impacts of climate change on marine fisheries, aquaculture, coastal tourism, and human health: An update. *Frontiers in Marine Science, 3*, 48. https://doi.org/10.3389/fmars.2016.00048.
80. Wittmann, A. C., & Pörtner, H. O. (2013). Sensitivities of extant animal ta□a to ocean acidification. *Nature Climate Change, 3*(11), 995–1001. https://doi.org/10.1038/nclimate1982.

81. Yazdi, S. K., & Shakouri, D. (2010). The effects of climate change on aquaculture. *International Journal of Environmental Science and Development, 1*(5), 378–382. https://doi.org/10.7763/IJESD.2010.V1.73.

82. Ytrestoyl, T., Aas, T. S., & Åsgård, T. (2015). Utilization of feed resources in production of Atlantic salmon (*Salmo salar*) in Norway. *Aquaculture, 448*, 365–374. https://doi.org/10.1016/j.aquaculture.2015.06.023.

83. Zhang, P. F., Zhao, T., Zhou, L., Han, G. D., Shen, Y. W., & Ke, C. H. (2019). Thermal tolerance traits of the undulated surf clam *Paphia undulata* based on heart rate and physiological energetics. *Aquaculture, 498*, 343–350. https://doi.org/10.1016/j.aquaculture.2018.08.037.

84. Zolnikov, T. R. (2018). Climate change. In T. R. Zolnikov (Ed.), *Autoethnographies on the Environment and Human Health* (pp. 32–37). Cham: Springer. https://doi.org/10.1007/978-3-319-69026-1.

85. Zolnikov, T. R. (2019). *Global Adaptation and Resilience to Climate Change*. Palgrave Studies in Climate Resilient Societies. Bern: Palgrave Macmillan. https://doi.org/10.1007/978-3-030-01213-7.

Chapter 6

Nanotechnology Applications in Aquaculture

Contents

INTRODUCTION

Aquaculture is regarded as a crucial food production system, with significant economic and food security implications. The continuous advancement of this industry plays a vital role in ensuring global nutritional safety. Currently, various nanotechnology-based systems are being used to enhance production, efficiency, and sustainability. Recent endeavors have focused on health management, improving the growth of fish and shellfish through the addition of nutraceuticals to their diet, as well as advancements in seafood processing, preservation, and water treatment. Hence, nanotechnology plays a crucial role in enhancing the efficiency and reducing the environmental footprint of this industry. From this standpoint, this chapter examines the present state of nanotechnology in the aquaculture and fisheries industry, focusing on both existing applications and future possibilities, as well as the ethical and governance considerations linked to this subject.

6.1. NANOTECHNOLOGY IN AQUACULTURE SYSTEMS

The United States National Nanotechnology Initiative (NNI) defines nanotechnology as the scientific comprehension and manipulation of matter at the nanoscale, specifically within the range of approximately 1 to 100 nm. This field explores the distinctive phenomena that occur at this scale, which in turn facilitate the development of innovative applications. Nanotechnology has significant potential to offer novel enhancements to aquaculture systems, resulting in cost reduction, improved efficiency, and reduced environmental impact. This is crucial in addressing the imperative of feeding the planet's population, which exceeds 7 billion. China has taken a leading role in the fast development and implementation of nanotechnology in the agri-food industry. The success of these applications is determined by their ability to meet the requirements of high quality, cost reduction, environmental sustainability, and minimal risk to human health (Chena & Yadab, 2011). Currently, nanotechnology is a rapidly growing industry worth billions of dollars, with over a thousand products containing nanomaterials available in the market. In the last ten years, more than 300 nanofood products have been introduced to global markets (Ramsden, 2018). The economic influence of nanotechnology industries is projected to reach at least one trillion dollars in 2020, and these industries are expected to provide employment for approximately six million individuals (He et al., 2019) (Figure 6.1).

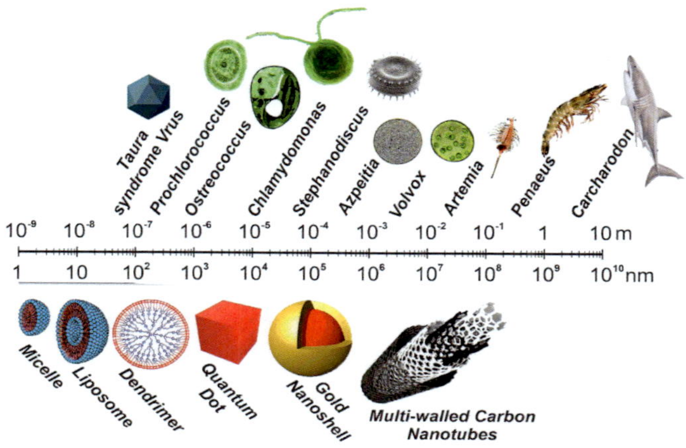

Figure 6.1. *Illustrates the proportional dimensions of nanoparticles in relation to various aquatic organisms.*

According to Technavio's 2019 report, the global nanotechnology market in the food sector is expected to grow at an annual rate of more than 24% between 2019 and 2023. This growth is primarily driven by the increasing use of nanotechnology in nutraceuticals. As a result, the market is projected to reach $112.48 billion by 2023. According to the existing data, the United States is currently the leading country in the nanofood sector, with China, Japan, and the European Union (EU) following closely behind (Chaudhry et al., 2017; Thiruvengadam et al., 2018). The United States is at the forefront of this group in terms of providing the largest funding for nanotechnology research. They have implemented a 4-year investment plan worth billions of dollars, which is administered through their National Nanotechnology Initiative (NNI) (Meghani et al., 2020; Thiruvengadam et al., 2018). Currently, prominent players in this sector include AQUANOVA (Germany), BASF (Germany), NanoPack (Belgium), and PEN (USA).

Nanotechnology encompasses a wide range of scientific fields, including physics, chemistry, biotechnology, and engineering (Chandra, 2016). It operates at various scales, where nanomaterials exhibit different structural elements such as crystallites, molecules, and clusters (Figure 6.2). These nanomaterials can be classified into different dimensions: zero dimension (nanoparticles, nanoclusters, and quantum dots), one dimension (carbon nanotubes and multiwalled nanotubes), two dimensions (graphene layers and ultrathin films), and three dimensions (nanostructured materials). Various types of nanoparticles are utilized, including dendrimers (Wu et al., 2015a), nanocapsules (Torchilin, 2006), nanospheres (Donbrow, 1991), nanotubes (Reilly, 2007), and others. In addition, nanomaterials have been found to possess numerous benefits, such as targeted delivery to specific tissues, decreased dosage and to□icity, enhanced bioavailability, improved drug effectiveness, and reduced secondary adverse effects (Shah & Mraz, 2019; Toyokawa et al., 2008; Xu et al., 2018a). The wide-ranging application of nanomaterials in areas such as food preservation, water treatment, and healthcare is due to their physiochemical properties (Baranwal et al., 2018; Ogunkalu, 2019).

Nanomaterials can be synthesized through either a bottom-up process, where atoms are arranged one by one, or a top-down process, where larger materials are converted into nanoscale. These processes are facilitated by the physical, optical, chemical, magnetic, and electric properties of nanomaterials (Roy et al., 2012). The primary categories of nanomaterials consist of nano-metals, metal oxides, carbon nanotubes, carbon spheres, and various composites like quantum dots, nano-ceramics, and nanoshells (Boxall et al., 2007; Stone et al., 2010).

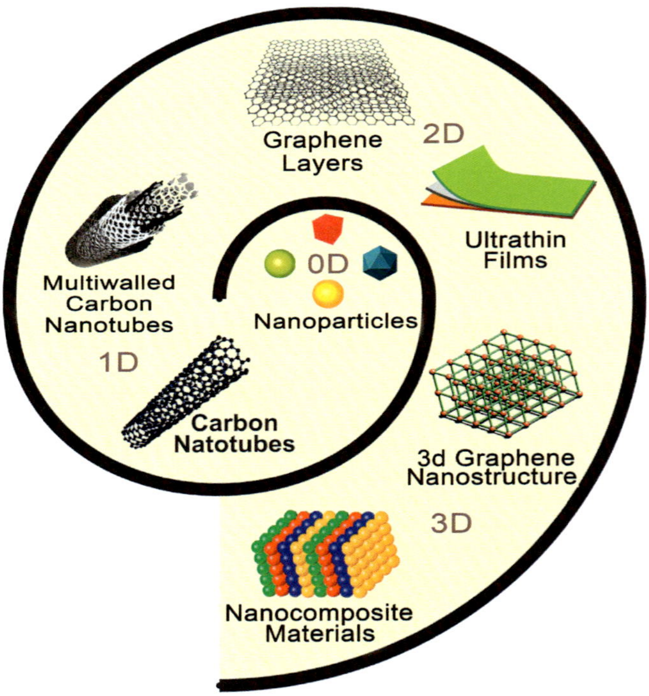

Figure 6.2. *Illustrates the various dimensions of nanomaterials that are presently under development.*

However, there is a vast range of chemical elements that have the potential to be utilized in the production of nanomaterials. Nanomaterials are currently being developed in intricate three-dimensional forms and may also consist of multiple chemical compounds (Handy, 2012, pp. 1-29). Conversely, the use of nanoparticles to enhance aquaculture and the seafood industry is rapidly gaining momentum. Aquaculture is the rapidly expanding sector of the food industry that accounts for the majority of seafood production for human consumption (Souza et al., 2017; Hussain et al., 2019). In 2018, global fish production amounted to appro□imately 180 million tons, with an estimated first sale value of billions of dollars. Out of this, aquaculture production contributed 82 million tons, valued at $250 billion. Over the years, per capita consumption of food fish has increased from 9.0 kg in 1961 to 20.5 kg in 2018, with an average annual growth rate of around 1.5% (FAO, 2020; Shah & Mraz, 2019). Aquaculture has the potential to make a significant contribution to achieving the Sustainable Development Goals for 2030, according to the Food and Agriculture Organization (FAO, 2018).

In addition, aquaculture plays a crucial role in providing employment opportunities for traditional fishing communities, particularly in developing nations. It currently employs approximately 20.5 million individuals globally (Sibaja et al., 2019; FAO, 2020). Environmental degradation, chemical contamination, suboptimal nutrition, and disease prevalence are factors that have a detrimental effect on the achievement of global goals in this sector.

Nanotechnology has diverse applications in aquaculture and has the potential to greatly enhance this industry (Figure 6.3). Among its present uses, we can identify the identification and management of disease-causing organisms, purification of water, disinfection of bodies of water, effective distribution of nutrients and medications (Jimenez-Fernandez et al., 2014; Bhattacharyya et al., 2015; Huang et al., 2015; Sibaja-Luis et al., 2019). For instance, DNA-nano vaccines are being utilized to enhance the immune system of fish. Furthermore, fish growth can be enhanced by utilizing iron nanoparticles (Mohammadi and Tukmechi, 2015).

Currently, there exist a multitude of potential applications for these technologies in the future. Research on the use of nanomaterials for delivery systems has been conducted within the agri-food sector. This research involves the utilization of various substances such as nanoparticles, micelles, liposomes, biopolymers, emulsions, proteincarbohydrate complexes, dendrimers, and solid nano-lipid particles. The key characteristics that offer benefits of these nanomaterials are their high absorption and bioavailability, improved dispersion and solubility, enhanced stability against environmental degradation during food processing, and controlled release kinetics (Chen et al., 2006; Ogunkalu, 2019; Pathakoti et al., 2017). In addition, the utilization of nanomaterials for delivery systems has the potential to enhance the nutritional characteristics of feed and improve the rate at which animals convert feed into body mass (Bhattacharyya et al., 2015). These benefits enhance efficiency, minimize waste and financial burden, and enhance production yield and quality (Chena & Yadab, 2011).

Nanotechnology applications offer a more efficient approach to delivering molecules for disease prevention and treatment compared to other technologies. This is achieved through the precise delivery and controlled release of drugs and vaccines, which helps to minimize health and environmental risks and reduce the reliance on chemicals. Nanoparticles used for targeted drug delivery offer the potential for faster, non-invasive, and cost-effective methods of drug administration. In addition, the integration of diagnostics and therapy into a single step, known as theragnostics, will enhance the efficacy of disease treatment and result in substantial cost reduction (Chena & Yadab, 2011; Morris, 2009).

Nanotechnology encompasses the advancement of fish packaging methods, as well as the enhancement of various aspects of fish quality such as flavor, te□ture, odor, appearance, and taste. Additionally, it aims to improve the absorption of nutrients in fish. Nanotechnology is utilized to enhance bioavailability, encompassing.

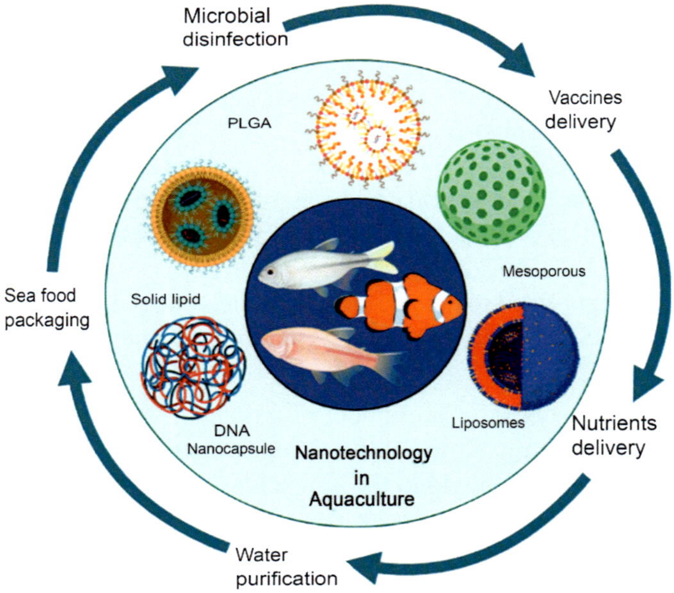

Figure 6.3. *Illustrates the main uses of nanotechnology in the aquaculture industry.*

The objectives of this chapter include the incorporation of functional compounds, the encapsulation and monitoring of the release of microencapsulated antimicrobials in packaging, the retardation of the decomposition process, and the enhancement of the stability and shelf-life of delicate ingredients. Another application of nanoparticles involves their bacteriostatic properties, which can be used to create surfaces that are resistant to microbial growth. For instance, silver nanoparticles have been employed in various sectors of the agri-food industry to reduce pathogens and prolong the shelf life of meat products (Ogunkalu, 2019).

However, the act of getting rid of nanoparticles also raises concerns regarding the environment and health (Otles & Yalcin, 2008). This is because the absorption of these particles can negatively impact the replication of DNA, leading to genetic mutations (Suppan, 2011, pp. 3-20). In order to ensure the long-term viability and expanded use of nanotechnology, it is necessary to thoroughly investigate the life cycle and shelf-life of nanomaterials. This research should focus on assessing the potential health and environmental risks associated with these materials, including their exposure, uptake, accumulation, release, and deposition (Handy, 2012, pp. 1-29).

6.2. DRUG DELIVERY FOR HEALTH MANAGEMENT

The occurrence of disease outbreaks poses significant challenges to the long-term viability and progress of aquaculture (Rather et al., 2011; Sibaja-Luis et al., 2019; Yang et al., 2021). Nanotechnology plays a significant role in this context, as it offers new perspectives on disease diagnosis and health management (Handy, 2012, pp. 1-29). One strategy involves using solid core drug delivery systems, where a solid nanoparticle is coated with a fatty acid shell to provide protection for the drug (Figure 6.4).

This method is particularly effective when dealing with drugs that are prone to change or are sensitive to temperature, as stated by Mitchell and Trivedi in 2010. Furthermore, porous nanomaterials can serve as a matrix for delivering pharmaceuticals. One example of using mesoporous silica particles is for the controlled release of drugs (Stromme et al., 2009). Oral nano-delivery systems connected to nanoparticles have been utilized for various purposes, such as improving the regulation of drug release (Eldridge et al., 1990), delivering drugs directly to target tissues (Jani et al., 1990), increasing the availability of pharmaceuticals with low absorption rates (Florence et al., 1995), prolonging the presence of drugs in the gut for stabilization purposes (Peters & Brain, 2009), and enhancing the absorption capacity through a higher rate of dispersion at the molecular level (Mohanraj & Chen, 2006).

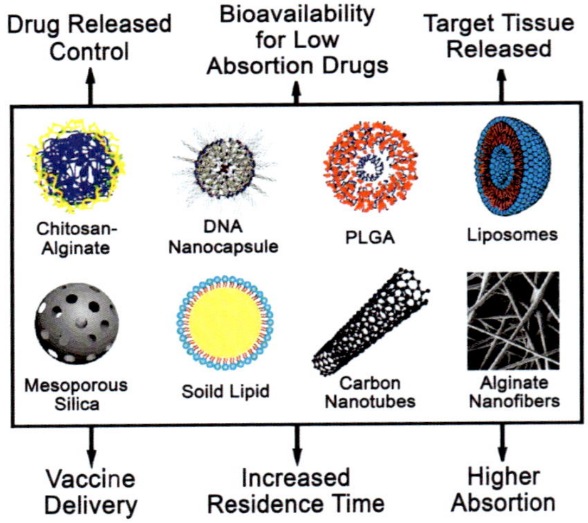

Figure 6.4. *Displays various nanoparticles currently under investigation in aquaculture for drug delivery and targeted therapy, along with the benefits associated with their use.*

Alginate is a biopolymer derived from brown algae and bacteria, consisting of -D-mannuronic acid (M) and -L-guluronic acid (G) (Shah & Mraz, 2019). It has the ability to produce nanoparticles through emulsification in a highly efficient and scalable manner (Reis et al., 2017). Multiple sources have reported that alginate serves as both an antigen adjuvant and a promoter of fish survival and weight (Borges et al., 2008; Tafaghodi et al., 2007; Chiu et al., 2008; Fujiki et al., 1994). In addition, alginate has been found to improve the immune response in the brown-marbled grouper (Epinephelus fuscoguttatus), the Asian carp (Cyprinus carpio), and the brown-marbled grouper and orange-spotted grouper (E. coioides) against iridovirus and Streptococcus sp. It has also been shown to enhance the immunological system of the turbot (Scophthalmus maximus) against V. anguillarum (Cheng et al., 2008; Huttenhuis et al., 2006; Skjermo & Bergh, 2004; Yeh et al., 2008). The combination of chitosan-alginate has proven to be highly effective in orally vaccinating rainbow trout (Oncorhynchus mykiss) against Lactococcus garvieae and Streptococcus iniae. This combination significantly improves survival rates and enhances the immune response when compared to the non-coated vaccine and control groups, as demonstrated in a study conducted by Halimi et al., in 2019.

In addition, DNA nanovaccines, which consist of small DNA strands enclosed in nanocapsules, have been employed in the aquaculture sector to stimulate an immune reaction in fish. The use of iron nanoparticles to expedite fish development through a drug delivery system with controlled release is rapidly becoming a reality, as demonstrated by Hussain et al., (2019). Nano encapsulated vaccines have effectively combated the bacterium Listonella anguillarum in Asian carp (Bhattacharyya et al., 2015) and rainbow trout (Oncorhynchus mykiss) (Mongillo, 2007; Ogunkalu, 2019).

Chitosan is a polysaccharide consisting of randomly distributed -linked d-glucosamine and -acetyl-d-glucosamine. It occurs naturally in the exoskeleton of crustaceans and insects (Elieh-Ali-Komi & Hamblin, 2016). It possesses distinctive characteristics, including non-toxicity, biodegradability, bio-adhesiveness, and biocompatibility. As a result, chitosan-based formulations, primarily in the form of emulsions, have been employed for drug delivery, edible coatings in aquaculture species, and various medical applications in humans, including surgical procedures and dentistry (Shah & Mraz, 2019).

There are numerous instances of effective encapsulation and transportation achieved through chitosan-based systems in aquaculture. Treatment methods have been developed to combat Vibrio parahaemolyticus in the blackhead seabream (Acanthopagrus schlegelii) (Li et al., 2013), Philasterides dicentrarchi in the turbot (S. ma□imus) (Leon-Rodrıguez et al., 2013), and Vibrio anguillarum in Asian sea bass (Lates calcarifer) (Kumar et al., 2008). Both DNA and RNA have been effectively encapsulated and delivered using this system. For instance, researchers have utilized dietary RNA in rohu (Labeo rohita) (Ferosekhan et al., 2014), and inactive particles of the viral haemorrhagic septicaemia virus (VHSV) in Japanese flounder (Paralichythys olivaceus) (Kole et al., 2019). In addition, Kitiyodom et al., (2019) have documented that immersion vaccination in tilapia (Oreochromis sp.) with a chitosan-coated mucoadhesive nanovaccine has resulted in improved effectiveness against columnaris disease. In a study conducted by Tandberg et al., in 2018, it was found that injecting chitosan-coated membrane vesicles (cMVs) from the intracellular fish pathogen Piscirickettsia salmonis into adult zebrafish (Danio rerio) resulted in a notable increase in survival rates and stimulated an enhanced immune response.

Rajeshkumar et al., (2009) discovered that chitosan-encapsulated DNA constructs containing the VP28 gene of white spot syndrome virus (WSSV) were effective in enhancing survival rates when exposed to the virus, in comparison to the mortality rate observed in the control group. Nevertheless, the relative survival rate decreased from 85% to an unspecified value when the challenge was conducted between 7 and 30 days after treatment. This decline can be attributed to the crustacean immune system's inability to generate a memory response.Polylactidecoglycotidic acid (PLGA), a biodegradable polymer, has been extensively employed for the encapsulation and delivery of various compounds in fish (Shah & Mraz, 2019). An instance of enhanced immunological response against lymphocystis was observed when a DNA vaccine was enclosed in PLGA, as demonstrated by Tian and Yu in 2011. In addition, Behera et al., (2010) discovered that the use of PLGA-encapsulated antigens from Aeromonas hydrophila as a vaccine in rohu fish resulted in a robust immune-stimulatory and antibody response, observed at both 21 and 42 days after immunization.

Yun et al., (2017) employed PLGA microparticles as a carrier for formalin-treated A. hydrophila cells, aiming to deliver antigens to pond loach (Misgurnus anguillicaudatus) and common carp (C. carpio). The fishes treated with PLGA-A. hydrophila exhibited enhanced survival rates and heightened innate and adaptive immune responses when faced with the challenge of A. hydrophila. This treatment has the potential to induce more robust and long-lasting immune responses compared to the group treated solely with killed A. hydrophila, as it combines the antigens with PLGA. In addition, mass vaccination can be accomplished by utilizing nanocapsules that are resistant to digestion and degradation. Both oral administration and targeted delivery of the active agent will decrease the effort and cost associated with disease management, resulting in more sustainable practices (Rather et al., 2011).

Liposomes, composed of phospholipids, have been extensively utilized in various areas of aquaculture. Regarding the Asian carp, an experiment conducted by Irie et al., (2005) demonstrated that the use of liposome-encapsulated antigens of Aeromonas salmonicida resulted in increased survival rates and reduced skin ulcers compared to the control group. In addition, in other experiments, the use of liposomes to emulsify A. hydrophila antigens resulted in an enhancement of serum antibody levels, thereby improving the immune response in common carp (Shah & Mraz, 2019; Yasumoto et al., 2006). In a recent study, Malheiros et al., (2020) discovered that nanoemulsions containing oleoresin from Copaifera reticulata (Leguminosae) can effectively combat monogenean parasites found on the gills of Tambaqui (Colossoma macropomum). Researchers discovered that the oleoresin, when not in nanoemulsion form, only demonstrated effectiveness at a concentration of or higher. However, when the oleoresin was in nanoemulsion form, it exhibited 100% efficacy after 30 minutes of e□posure at a concentration of . In addition, all fish perished within 2 hours of being e□posed to all tested concentrations (200-1000 mg/L) when applied individually. Conversely, all fish treated with nanoemulsioned oleoresin remained alive across all tested concentrations (50-250 mg/L).

Recently, there have been advancements in the development of controlled release delivery systems and diagnostic sensors that utilize nanoparticles. These nanoparticles are designed to alter their properties and structure in response to various environmental factors, such as temperature, ionic strength, pH, or enzymatic activity. In the pharmaceutical industry, pH sensitive nanoparticles have been utilized to transport anticancer molecules (Nile et al., 2020). In addition, McClements (2017) documented that lipid nanoparticles containing bioactive compounds can be enzymatically degraded by lipases, resulting in the release of the bioactive compound in cases of gastrointestinal disorders. In a recent study, Lee et al., (2020) devised a compelling technique with potential applications as oral vaccines in aquaculture. In order to investigate their methodology, the researchers employed an immune system stimulator, specifically the antigen derived from Streptococcus parauberis that had been rendered inactive through formalin treatment. This antigen was then enclosed within alginate beads and served as a model for their study. To regulate both the efficiency of encapsulation and the release at the specific organ (intestine), they interconnected the beads with clay nanoparticles. The entire system is designed to store substances in the stomach and release them in the intestines, in response to changes in pH levels. This approach has the potential to be highly effective in selectively treating only the affected organisms within a cultured population, employing the treatment solely when necessary.

6.3. NANOSENSORS FOR PATHOGEN DETECTION

Researchers are currently working on developing nanobiosensor systems that can detect extremely low concentrations of parasites, bacteria, viruses, and pollutants in water (Chen et al., 2016).

Timely identification of the etiological agent in outbreaks at commercial aquaculture systems is crucial. Delaying the treatment to control the pathogen can have significant economic consequences, as the impact of the etiological agent may not be immediately apparent. Nanotechnology has the potential to address this challenge by detecting and eliminating pathogens at an early stage.

At present, nanosensors have the capability to identify a diverse array of disease-causing agents. Using electrical nanosensors, it is possible to detect individual virus particles (Patolsky et al., 2004). Furthermore, studies have shown that gold nanoparticles can be modified with an antibody that specifically targets a specific biomolecule, such as immunoglobulin G-capped gold nanoparticles, which can bind specifically to antibodies produced against Staphylococcus pyrogenes, S. aureus, and other similar bacteria (Roy et al., 2012). Nanosensors, specifically those utilizing carbon nanotubes, are employed in the maintenance of fishponds and the monitoring of livestock. These nanosensors exhibit exceptional sensitivity in detecting minute quantities of pathogens such as viruses, parasites, and bacteria, as well as heavy metals present in both food and water sources (Hussain et al., 2019).

Researchers have reported the use of tracking nanosensors equipped with locators to collect and analyze data on the geolocalization and health status of fish. This technology enables the monitoring and control of individual fish as well as the development of intelligent cage systems. (Sekhon, 2014; Hussain et al., 2019). A chip containing a nanoscale radio circuit and an embedded identification code, known as a nano-barcode, has been developed as a tracking device. This chip allows for automatic identification of the tagged items from a distance by scanning the information stored in the tags. These tags have the potential to serve as both a tracking device and a means to monitor various aspects of animal behavior, including feeding patterns, swimming behavior, and metabolic activity. Nano-barcoding enables exporters and the processing industry to effectively monitor the origin of a specific product and track its delivery status. In addition, these nanosensors, when combined with synthetic DNA labeled with color-coded probes, nano-barcode systems, can be utilized not only for pathogen identification but also for monitoring leakages, temperature fluctuations, and other parameters. This, in turn, improves the quality of the product (Rather et al., 2011).

6.4. MICROBIAL DISINFECTION

Several metallic nanoparticles, including silver, titanium, and copper, have been utilized for the purpose of disease prevention and treatment. Metal nanoparticles exhibit various mechanisms of action against bacteria, with one particularly potent effect being their ability to attach to the cell membrane and cell wall through electrostatic interactions, leading to their disruption (Figure 6.5).

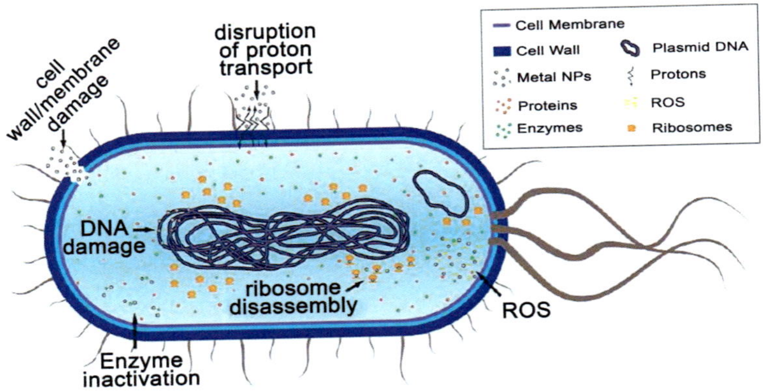

Figure 6.5. *Illustrates the impact of metal nanoparticles (NPs) on various components of bacteria during their application for microbial disinfection, disease treatment, and prevention.*

Metal nanoparticles have the ability to attach to and disturb both the cell membrane and the cell wall. Additionally, they can interfere with ion transport by binding to ions and ion channels. These nanoparticles (NPs) have the ability to induce double strand breaks in the DNA, disrupt the assembly of ribosomes, and interfere with enzymatic activity through electrostatic interactions. Metal nanoparticles can also induce a heightened state of oxidative stress, leading to an increase in the production of reactive oxygen species (ROS). These ROS can cause damage to proteins, lipids, and DNA. In addition, they have the ability to interfere with ion transport through their interaction with ions and ion channels. Nanoparticles (NPs) can induce double strand breaks in DNA, disrupt ribosome assembly, and interfere with enzymatic activity through electrostatic interactions. Metal nanoparticles are also recognized for inducing an elevated state of oxidative stress, leading to an increased production of reactive oxygen species (ROS) that have the potential to harm proteins, lipids, and DNA.

Colloidal silver nanoparticles are a prominent nanotechnology product utilized to combat a broad range of pathogens, encompassing viruses, parasites, fungi, and bacteria. Silver is highly regarded as a promising nanomaterial among oligo-dynamic metallic nanoparticles. This is because it has a wide range of effects on various microbial species, is easy to apply in different forms, has a crystallographic structure, a high surface area compared to its volume, and is compatible with several compounds (Nangmenyi & Economy, 2014). The antimicrobial activity is associated with its ability to oxidize DNA and proteins, resulting in significant damage. For instance, silver nanoparticles have the ability to eradicate various types of methicillin-resistant S. aureus bacteria (Jeong et al., 2005).

Visible light photo-catalysts mediated by metal oxide nanoparticles, as well as nano-porous fibers and foams, have been developed for microbial elimination protocols (Li et al., 2014). Their impact extends beyond the eradication of microbes to include the removal of organic pollutants from the pharmaceutical and cosmetics sectors (Chena & Yadab, 2011). Titanium dioxide nanoparticles e□hibit potent antibacterial properties, effectively eliminating fish pathogens in laboratory conditions (Cheng et al., 2009). Additionally, they have been shown to significantly influence the immune response of fish neutrophils, acting as a robust immune modulator (Jovanovic et al., 2011). Presently, there exists a nanotechnology device called NanoCheck that is commercially accessible. This device utilizes lanthanum based particles, measuring 40 nm in size, to effectively clean fishponds. The particles function by absorbing phosphates from the water, thereby inhibiting the growth of algae (Rather et al., 2011).

In their study, Korni and Khalil (2017) found that ginger nanoparticles have the capability to inhibit the occurrence of motile Aeromonas septicaemia infection in Asian carp fingerlings. Rather et al., (2017) found that silver nanoparticles produced by Azadirachta indica (neem) have the ability to inhibit the growth of bacteria and modulate the immune response in mrigal (Cirrhinus mrigala) fingerlings when e□posed to A. hydrophila. Erdem et al., (2018) recently showed that silver nanoparticles synthesized through Aeromonas sobria have antibacterial effects against A. hydrophila. This has opened up the potential for enhancing sanitary management by utilizing antimicrobial agents that are more ecologically sustainable (Shah & Mraz, 2019).

It is noteworthy to mention the recent publication by Kepiro et al., (2020) that presents the conceptual design of protein pseudocapsids with a wide range of antimicrobial properties. Unlike traditional antibiotics, these pseudocapsids e□hibit e□ceptional efficacy against a wide range of bacterial strains. This technique has the ability to eradicate antibiotic-resistant bacteria inside a living organism without inducing any harmful effects. Pseudocapsids exhibit an icosahedral structure that can vary in size but not in shape. They are present in both D and L epimeric configurations. The authors provide evidence that these pseudocapsids can cause rapid and irreversible damage to bacterial cells.

6.5. TREATMENT OF POLLUTANTS IN THE WATER

Nanotechnology has been utilized to tackle water pollution, a significant problem in aquaculture. The use of nanomaterials in water treatment allows for the creation of cost-efficient and highly effective techniques for purifying water, particularly through photo-catalysis and adsorption processes. A technique employed to eliminate arsenite pollution from groundwater involves the creation of magnetic konjac glucomannan aerogels possessing environmentally sustainable properties (Ye et al., 2016). Graphene nano-sheets and graphene oxide have garnered considerable attention in recent years due to their capacity to eliminate diverse contaminants from water (Liu et al., 2016; Motamedi et al., 2014). Graphene oxide-titanium oxide nanocomposites have been used to adsorb and remove heavy metals and organic compounds from residual water (Atchudan et al., 2017; Hu et al., 2013). Titanium oxide exhibits numerous advantageous properties, such as cost-effectiveness, non-toxicity, exceptional photo-catalytic efficacy, and robustness in both biological and chemical settings. These characteristics render it a highly auspicious choice for the treatment of residual water. Furthermore, a multitude of studies have been undertaken to examine the photocatalytic characteristics of titanium dioxide. The studies have shown that it can effectively eradicate different types of bacteria (both Gram-negative and Gram-positive), fungi (both filamentous and unicellular), algae, protozoa, mammalian viruses, and bacteriophages (Foster et al., 2011). The activity revolves around the production of reactive oxygen radicals and peroxides, which possess the capacity to harm cell walls and membranes. Nanoparticles possess the capacity to degrade antibiotics by producing reactive oxygen species. The utilization of iron nanoparticles was utilized to break down polychlorinated biphenyls and dioxins into less harmful carbon compounds in groundwater, employing comparable techniques (Majumder & Dash, 2017).

6.6. DELIVERY OF DIETARY SUPPLEMENTS AND NUTRACEUTICALS

The primary underlying principle behind the notion that nanoparticles can enhance fish development is their capacity to augment the absorption of nutrients throughout the digestive tract. Nanoparticles of micronutrients, when added to aquaculture feeds, can more effectively enter cells, resulting in an increased absorption rate (Figure 6.6). (Ogunkalu, 2019; Zhou et al., 2009). Iron nanoparticles were found to enhance the growth rates of sturgeon and young carp, as demonstrated in a study conducted by ETC in 2003. In the case of nano-selenium, it was found that it was more effective than organic seleno-methionine, yielding a similar outcome. The addition of nano-selenium to diets can increase the concentration of selenium in the muscles, improve antioxidant levels, boost the rate of weight gain, and ultimately increase the final weight of crucian carp (Carassius auratus gibelio) (Zhou et al., 2009).

The addition of selenium nanoparticles to the feed leads to a notable enhancement in both the antioxidant defense system and growth (Ashouri et al., 2015). Furthermore, a study conducted by Izquierdo et al., (2017) found that adding manganese, zinc, and selenium nanoparticles to the early diets of gilthead seabream (Sparus aurata) resulted in improved bone mineralization and increased resistance to stress. The addition of Lactobacillus casei and iron nanoparticles to the diet of rainbow trout significantly improved their growth parameters (Mohammadi et al., 2015). Similarly, the inclusion of manganese oxide nanoparticles in the diet of freshwater prawn (Macrobrachium rosenbergii) greatly enhanced their antioxidant defense system and growth (Asaikkutti et al., 2016). In addition, the inclusion of copper nanoparticles in the feed significantly enhanced the non-specific immune response, levels of antio□idant metabolic enzymes, activities of digestive enzymes, biochemical constituents, and growth in both red sea bream (Pagrus major) (ElBasuini et al., 2017) and freshwater prawn (Muralisankar et al., 2016).

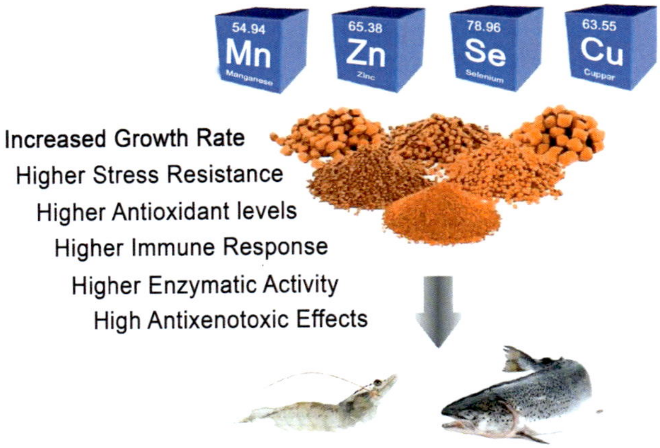

Figure 6.6. *Depicts the use of microelements in the form of nanoparticles to be included in aquaculture feeds and the effects seen in shrimp and fish.*

In a study conducted by Kunjiappan et al., (2015), it was found that gold nanoparticles synthesized by Azolla microphylla using phytochemical methods can effectively reduce hepatic damage in Asian carp caused by acetaminophen contamination in the water. The gold nanoparticles significantly enhance the levels of oxidative stress markers, decrease hepatic ions, metabolic enzymes, hepatotoxic markers, abnormal liver histology, and altered tissue enzymes. According to Sharif et al., (2017), Aloe vera nanoparticles have been found to improve body composition, survival rate, and growth in the Siberian sturgeon (Acipenser baerii).

Nanotechnology has been used to enhance the bioavailability and prolong the retention time of natural bioactive compounds (Cui et al., 2009). For instance, researchers have reported the encapsulation of curcumin in nanoparticles using phospholipids (Semalty et al., 2010), micelles (Takahashi et al., 2009; Wang et al., 2009), liposomes (Letchford et al., 2008), hydrogels (Shah et al., 2008), and other methods (Das et al., 2010). In a recent study by Shah et al., (2016), curcumin was successfully encapsulated in chitosan nanoparticles using Pickering emulsion, leading to an enhanced shelf-life of curcumin. When it comes to hydrophobic bioactive compounds, the Pickering emulsion is regarded as the most effective method for encapsulation. This approach enhances the shelf-life of the compounds by using solid particles (such as nutrients) to stabilize them (Matos et al., 2016; Shah & Mraz, 2019; Xu et al., 2018b).

These emulsions exhibit superior properties compared to traditional emulsions (Dickinson, 2010), including low or no toxicity, improved reproducibility and biocompatibility, easy and scalable production (Wu & Ma, 2016), and enhanced stability (Binks, 2007). In addition, the composition and structure of these emulsions enhance the transportation of biological substances in the digestive system (Liu & Tang, 2016; Matos et al., 2018; Ngwabebhoh et al., 2018; Shao et al., 2018; Winuprasith et al., 2018). Presently, a wide range of colloidal particles are employed in the manufacturing of Pickering emulsions, wherein solid particles aid in stabilizing the emulsion. Starch-based Pickering emulsions have been employed as a vehicle for delivering compounds, such as hydrophobic antifungal compounds, as demonstrated by Marku et al., (2012), Cossu et al., (2015), and Leclercq & Nardello-Rataj (2016).

The utilization of nano-crystalline, self-stabilized Pickering emulsion shows significant promise for drug delivery and release, particularly for compounds with limited solubility (Yi et al., 2017). In a recent study by Zhou et al., (2018), cellulose nanocrystals were used to stabilize oil Pickering emulsions containing oregano essential oil. The purpose of this was to enhance the antibacterial properties of the emulsions against E. coli, B. subtilis, S. aureus, and S. cerevisiae. Furthermore, to create a method of delivering antibacterial agents specifically targeting E. coli, we utilized Zein/Arabic-gum nanoparticles for stabilization. A Pickering emulsion was formed using thymol as reported by Li et al., (2018) and Shah & Mraz (2019). Recently, Baldissera et al., (2020a; b) discovered that adding nerolidol-loaded nanospheres to the diet of Nile tilapia (Oreochromis niloticus) infected with Streptococcus agalactiae resulted in decreased bacterial levels, improved survival rates, and protection against oxidative damage.

Furthermore, nanoparticles can also be utilized to enhance the stability and bioavailability of food ingredients, as well as to alter the physical characteristics of fish food. Even minute additions of nanomaterials can significantly improve the physical characteristics of food pellets. For instance, when single-walled carbon nanotubes are added to the diets of trout, it leads to the formation of a solid pellet that remains intact when submerged in water. It is crucial to decrease pollution and food wastage in aquaculture systems caused by improper buoyancy, inadequate food stability, or unsatisfactory texture of the pellets, which result in substantial losses in this industry (Handy & Poxton, 1993). Currently, there is significant research and investigation being conducted in the industry regarding the advancement of nano-formulations. An inherent attribute of these systems is their versatility in various applications, including the administration of antibiotics, vaccines, pharmaceuticals, and nutraceuticals, among others (Rather et al., 2011; Sibaja-Luis et al., 2019).

Recent studies have concentrated on investigating various biopolymers that can be utilized in the aquaculture industry (Alboofetileh et al., 2016; Borgogna et al., 2011; Dursun et al., 2010; Joukar et al., 2017). A study conducted by Alishahi et al., (2011) discovered that when vitamin C is enclosed in nanoparticles made from chitosan, it leads to an increase in the levels of vitamin C in the serum of rainbow trout (*O. mykiss*) when it is included in their diet. This increase in vitamin C levels also enhances the innate immune response, as indicated by higher levels of lysozyme and hemolytic serum complement activities, when compared to control groups that were given vitamin C and chitosan separately.

In a recent study conducted by Abd El-Naby et al., (2019), it was found that adding chitosan nanoparticles to the diet of *Nile tilapia* (*O. niloticus*) can enhance both the efficiency of feed utilization and the growth of the fish. According to these authors, the presence of chitosan particles enhanced the activity of enzymes like lipase and amylase. Furthermore, it has been discovered that the utilization of these nanoparticles not only hinders the proliferation of both aerobic and anaerobic bacteria, but also enhances the innate immune system. In addition, Naiel et al., (2020) have documented that the combination of vitamin C with chitosan nanoparticles exhibits an immunomodulatory effect against the toxicity induced by exposure to the pesticide imidacloprid. Additionally, they highlight that the addition of this mixture leads to enhanced growth and utilization of food in the presence of water-borne pesticides. Furthermore, they highlight that the incorporation of chitosan nanoparticles and vitamin C delivered through food has not only a beneficial impact on the antio□idant state and non-specific immune response, but it may also have additional favorable health effects. These effects can be observed in the structure of hepatocytes and the overall health condition of . niloticus when exposed to sub-lethal levels of imidacloprid. In a recent study conducted by Ismael et al., (2021), it was found that the addition of dietary zeolites, either alone or in conjunction with chitosan nanoparticles, can effectively reduce the harmful effects of imidacloprid toxicity. However, Abd El-Naby et al., (2020) suggest that the combination of chitosan nanoparticles with thymol can enhance the growth and utilization of food in . niloticus. This combination not only increases the activity of lipases, catalases, and proteases, but also promotes the lengthening of the intestinal villus. However, the use of nanomaterials as nutraceuticals in aquaculture (specifically shellfish and fish) is currently in the early stages of consideration. These nutraceuticals are being explored for their potential to enhance the value of aquaculture products, reduce stress, and manage health. Despite their potential benefits, the use of nutraceuticals remains limited due to their high cost (Ogunkalu, 2019).

6.7. SEAFOOD PROCESSING

Nanotechnology has the potential to be applied in the manufacturing of aquaculture species and in the promotion of seafood, primarily due to the need to prolong the shelf-life of the products. Research in the food packaging industry has been intensified through the use of nanomaterials. These materials offer new properties such as oxygen depletion, reduction of enzyme activity and product degradation, as well as antimicrobial and antifungal activities. They also enable the detection of pathogens and toxins, ultimately improving the stability of food products (Jiang et al., 2015; Kumar et al., 2020; Kuswandi, 2016; Mihindukulasuriya & Lim, 2014; Reig et al., 2014; Rhim et al., 2013; Sibaja-Luis et al., 2019; Siegrist et al., 2008).

An essential objective of the food industry is to prolong the duration that a product can be stored, with the aim of maintaining both its freshness and overall quality. At present, nanotechnology has the capability to produce significant advancements in the field of food processing. These advancements are aimed at extending the shelf life of products by preventing the movement of gases through the packaging. Additionally, nanotechnology can provide information about potential spoilage of food components (Nickols-Richardson & Piehowski, 2008). Nanostructures such as nanoemulsion, nanofibers, and nanoparticles have the ability to retard the deterioration in quality, thereby preserving the color and flavor of the product (Chellaram et al., 2014; Ozogul et al., 2017).

Conversely, there have been numerous reports on the utilization of essential oils enclosed within various nanostructures, such as cyclic oligosaccharides (Abarca et al., 2016; Ciobanu et al., 2012; Hill et al., 2013; Siqueira-Lima et al., 2014), nanotubes (Kim et al., 2016; Lee & Park, 2015), polymeric nanoparticles (Christofoli et al., 2015; De Oliveira et al., 2014; Liakos et al., 2016), solid lipid nanoparticles (Cortes-Rojas et al., 2014; Feng, 2012; Lai et al., 2006; Moghimipour et al., 2013), zein nanoparticles (Parris et al., 2005; Wu et al., 2012; Zhang et al., 2014; da Rosa et al., 2015), and biodegradable nanoparticles (Chifiriuc et al., 2017; Pavela et al., 2017; Sotelo-Boyas et al., 2017a,b), etc.

In addition, ice infused with nanoparticles can be utilized for various food packaging purposes. For instance, researchers have found that silver nanoparticles derived from banana midrib and incorporated into nano-ice can effectively decrease the presence of microorganisms on the surface of flathead grey mullet (Mugil cephalus), and even prevent the growth of Acinetobacter bacteria (Daniel et al., 2016). Organic methods for producing silver nanoparticles are receiving more attention in the field of seafood preservation compared to chemical synthesis (Huang et al., 2018). Furthermore, certain organisms, such as oysters, prawns, and fish, have also been utilized for the synthesis of nanoparticles. These organisms have a rich supply of bioactive compounds, including minerals, oils, proteins, lipids, flavonoids, vitamins, polyphenols, fibers, polysaccharides (fucoidan, laminaran, and alginate), terpenoids, and carotenoids. These compounds have various potential ethnobotanical functions (Kushnerova et al., 2010; Hussain et al., 2019).

6.7.1. Nanofilms

Nanotechnology is being used in the preservation of seafood to delay the growth of microorganisms and the breakdown of enzymes, thereby enhancing food security in the field of conservation and packaging techniques. There is a growing popularity of products that utilize nanomaterials with antimicrobial properties, such as packaging materials that contain silver nanoparticles (Vasile, 2018; Hussain et al., 2019). Nanocomposite films are added to food products in conjunction with antimicrobial films and edible coating techniques. Nanocomposite films are produced from naturally occurring biopolymers such as lipids, proteins, and polysaccharides. Due to their environmentally friendly, anti-carcinogenic properties, and ability to be consumed, these alternative packaging materials are being increasingly used to replace plastics made from petrochemical sources (Dursun et al., 2010; Ogunkalu, 2019). The study conducted by O'Callaghan and Kerry (2016) examined various types of chitosan nanoparticles that have antimicrobial properties. Nanocomposite films made from chitosan nanoparticles and gelatin, along with oregano essential oil, exhibit strong antimicrobial properties. These films are effective against the four most prevalent foodborne pathogens, namely E. coli, Salmonella enteritidis, S. aureus, and Listeria monocytogenes (Hosseini et al., 2016; Hussain et al., 2019).

Recent research suggests that chitosan, a natural substance, shows promise as an edible coating material for sea-products. It has the potential to prolong shelf life and enhance the microbiological quality of these products (Mohan et al., 2012). Chitosan is regarded as an effective antimicrobial agent due to its poly cationic form (Suptijah et al., 2008). In addition, chitosan nanoparticles possess remarkable physiochemical properties, such as bioactivity (Yang et al., 2010). Several studies indicate that incorporating nanoparticles into bio-nanocomposite films, specifically chitosan nanoparticles and gelatin, enhances their barrier properties (Hosseini et al., 2015). Furthermore, it has been confirmed that nano-chitosan is a superior antibacterial agent in comparison to chitosan coating applied on silver carp fillets (Hypophthalmichthys molitri□). Chitosan has been extensively researched as a preservation agent for sea-products in Indonesia, with a focus on pond-raised species such as Nile tilapia (Oreochromis sp.) (Tapilatu et al., 2016) and catfish (Pangasiusypopthalmus) (Suptijah et al., 2008). Furthermore, there have been reports of the use of chitosan in tropical marine fishes, particularly in pre-treated products like salted Indian scads (Decapterus sp.) (Swastawati et al., 2008; Hussain et al., 2019).

Carbon nanotubes, fullerene, graphene, and graphene oxide possess excellent antimicrobial properties. They cause physical harm by damaging the cell wall and membrane, and also induce chemical damage by oxidizing phospholipids through the generation of reactive oxygen species (Azizi-Lalabadi et al., 2020). These nanomaterials have significant potential for the development of new nanocomposite materials in the seafood packaging industry.

6.7.2. Nanoemulsions

Nano-emulsions exhibit exceptional physicochemical properties and stability due to their extremely small droplet size (McClements, 2011; Otoni et al., 2016; Solans et al., 2005). These dimensions result in a greater surface area relative to the mass, thereby enhancing the effectiveness of lipophilic solutions (McClements & Rao, 2011). Nano-emulsions are at the forefront of advancements in producing edible films. Researchers (Acevedo-Fani et al., 2015) have developed alginate films containing nano-emulsions loaded with essential oils. This was done to enhance water dispersion and protect the essential oils from degradation. The use of nano-emulsion containing essential oils, combined with the antibacterial properties of manufactured films, has been shown to inhibit the growth of E. coli. Additionally, an edible composite film made with nano-emulsified cinnamaldehyde has demonstrated antimicrobial activity. These findings have been reported by Otoni et al., (2014) and Hussain et al., (2019).

6.7.3. Nanoliposomes

Essential oils have demonstrated their superior and enhanced antimicrobial abilities in surfactant nano-metric micelles. Nano-liposomes containing essential oils exhibit controlled release, thereby enhancing the antibacterial efficacy of the film. As an illustration, Wu et al., (2015b) created nano-liposomes containing cinnamon essential oil, which were then added to gelatin films. This resulted in a decrease in the rate of release and an increase in the effectiveness of the antimicrobial properties. In addition, the process of encapsulating carvacrol and eugenol into nano-sized surfactant micelles results in a significant antimicrobial effect (Gaysinsky et al., 2005). The researchers in Jimenez et al., (2014) achieved a highly stable form of nano-liposomes by using soy lecithin and rapeseed to encapsulate orange essential oil. These nano-liposomes were then added to starch films made of caseinate. Valencia-Sullca et al., (2016) demonstrated the capacity of lecithin nano-liposomes to encapsulate essential oils. In addition, the utilization of eugenol in nano-liposomes containing lecithin enables the films to preserve approximately 40-50% of the eugenol that is incorporated. This is in contrast to a direct emulsification method, which only retains 1-2% of the compound (Hussain et al., 2019).

6.8. NANOTECHNOLOGY RISK ASSESSMENT AND GOVERNANCE

Presently, there is growing public concern regarding the potential toxicity of nanoparticles in biological systems. Nanomaterials have the potential to become a new type of pollutants in the environment. Current research is primarily concerned with investigating the potential adverse effects they may cause (Moore, 2006; Shah & Mraz, 2019). Due to their minuscule dimensions, nanoparticles are capable of infiltrating cell membranes and inducing genotoxicity. The inherent chemical reactivity of nanomaterials leads to an increased generation of reactive oxygen species and free radicals, which is a primary mechanism of toxicity for nanoparticles. This can result in not only inflammation and oxidative stress, but also harm to proteins and DNA. Studies have shown that nanomaterials have the capacity to cause DNA mutations and significant structural harm to mitochondria, which may lead to cell death (Majumder & Dash, 2017; Meghani et al., 2020; Vicari et al., 2018).

Multiple studies have documented the harmful effects of nanomaterials on aquatic organisms (Asharani et al., 2008; Griffitt et al., 2008; Matranga & Corsi, 2012; Garcia-Negrete et al., 2013; Canesi et al., 2015; Vignardi et al., 2015; Swain et al., 2016; Canesi et al., 2017; Sendra et al., 2017a; b; c; 2018a; b; 2019; Aouini et al., 2018; Volland et al., 2018; Naguib et al., 2020; Sayed et al., 2020). Studies in nanotoxicology have been carried out to elucidate both the regulatory aspects and bio-distribution of these components, as well as to assess their potential risks to the environment at the molecular level (Bello & Leong, 2017). For the fisheries and aquaculture industries to effectively incorporate nanotechnology in a sustainable manner, further research is needed to investigate the potential accumulation of nanomaterials and their potential toxicity to aquatic organisms and humans (Skjolding et al., 2016; Sibaja-Luis et al., 2019). Enhancing public engagement is essential for maintaining trust in nanotechnology, particularly regarding the safety of food and the environment.

A crucial inquiry persists regarding public sentiment and the willingness of consumers to consume fish that may potentially contain nanoparticles. The public perception of nanotechnology appears to be significantly more positive compared to genetically modified organisms (GMOs). This can be partly attributed to the industry's proactive efforts in educating the general public about the development of nanomaterials, starting from their early stages (Anderson et al., 2005; Handy, 2012, pp. 1-29; Handy & Shaw, 2007; He et al., 2019; Mukherjee et al., 2019). Additionally, further risk assessments regarding the potential long-term consequences of incorporating engineered nanomaterials into human food should be conducted. Regarding the aquaculture sector, the primary risk of exposure is associated with the consumable portions of the fish, such as the muscle tissue. Current data indicate that the levels of metal nanoparticles in fish tissues, resulting from the consumption of diets containing these nanoparticles, are either in the same range or lower. This information is supported by studies conducted by Handy et al., (2011), Khosravi-Katuli et al., (2017), Ramsden et al., (2009), and Shaw & Handy (2011). Therefore, potential health hazards for humans are expected to be linked to the long-term impacts caused by consuming fish that may contain nanomaterials in amounts equal to or below the ng level. Nevertheless, these types of concerns are not novel. The legislation on allowable levels of pollutants in aquaculture regulates the acceptable risk posed by persistent chemicals, such as mercury or pesticides, in edible fish (Berntssen et al., 2010). Analogously, the identical framework can be applied to nanomaterials. Furthermore, it would be wise to conduct further research on the public's willingness to accept nanomaterials in edible seafood products. Hence, it is imperative to exert additional endeavor in achieving public engagement. This is necessary to elucidate the present applications of nanotechnology in this sector, as well as to communicate the advantages and potential hazards to the consumers (Handy, 2012, pp. 1-29).

We should not let such concerns hinder our efforts to advance nanotechnology in the aquaculture industry. Implementing a meticulous surveillance system combined with regulated utilization can assist in the endeavor to enhance advantages and mitigate hazards (Ashraf et al., 2011). A safety handling guide is provided for laboratory workers who use nanomaterials for research purposes (Handy, 2012, pp. 1-29). Currently, there is a lack of standardized protocols for assessing the impacts of artificial nanomaterials on the environment. This is primarily because of the extensive variety of nanomaterials and the intricate nature of their dispersion dynamics in the environment. Therefore, assessments of environmental levels primarily rely on conceptual models of nanoparticle emission. The majority of the knowledge regarding the behavior and destiny of nanomaterials is derived from research conducted in freshwater ecosystems (Handy et al., 2008; Ju-Nam & Lead, 2008; Klaine et al., 2008), as well as in marine environments (Bundschuh et al., 2018; Canesi & Corsi, 2016; Selck et al., 2016). The concentration levels of artificially produced nanomaterials in surface waters were estimated to be in the range of nanograms or lower, based on various models (Boxall et al., 2007; Bundschuh et al., 2018; Gottschalk et al., 2009; Metreveli et al., 2016; Nowack & Bucheli, 2007). Recent projections for freshwater nanomaterial concentrations vary from a few nanograms per liter in 2017 to several hundred nanograms per liter in 2050. When comparing with regular materials, the estimates for -nanomaterials are almost one thousand times higher, whereas for silver-nanomaterials they are ten times lower. Nanomaterials generally pose a low risk to most environmental niches. However, organisms residing in close proximity to nanomaterial sources, such as waste treatment facilities or production plant outfalls, may face a higher risk (Giese et al., 2018). These predictions are speculative, and our understanding of how manufactured nanomaterials behave in crucial ecosystems, such as the oceans, is currently limited. However, Handy (2012) and Shah & Mraz (2019) provided us with valuable information regarding the potential eco-toxicity, behavior, and fate of synthetic nanomaterials in marine environments. They also discussed various surveillance and monitoring strategies, as highlighted by Abdolahpur et al., (2019), Baalousha et al., (2016), Cerrillo et al., (2017), and Sun et al., (2016).

The potential risks associated with nanomaterials, like any novel substance, should be carefully weighed against the potential benefits (Owen & Handy, 2007). The evaluation of risks, advantages, and ethical concerns is contingent upon multiple factors, which encompass not only the composition of nanomaterials but also the particular field of implementation, deployment techniques, and ultimate objectives. In recent decades, significant attention has been focused on legislation, which has undergone extensive evaluation (Jain et al., 2018; Kaphle et al., 2018; Wacker et al., 2016). The consensus reached is that it is necessary to develop experimental protocols to assess the potential effects of a material before its use in agriculture, food, and the environment. This will enable the creation of regulatory guidelines that can fully exploit the benefits of nanotechnology.

Currently, there are no established regulatory laws governing the utilization of nanotechnology in the food and agricultural industry. Therefore, it is imperative to establish efficient policies and guidelines to ensure the safe utilization of nanoparticles in the industry. As a result of these constraints, the current legislation is still in its early stages of progress, focusing only on broad strategies pertaining to nanomaterials and nanotechnology. However, there is currently no global standardized regulation for synthetic nanomaterials. Products that are not covered by the classification and labeling requirements of the United Nations' Globally Harmonized System of Classification and Labelling of Chemicals (GHS) do not provide any information for consumers in the form of declaration statements. The management of nano-waste was subject to the same situation (Karlaganis et al., 2019).

For instance, the Food and Drug Administration (FDA) permits the direct consumption of inorganic oxide metals like titanium dioxide (E171), silicon dioxide (E551), and magnesium oxide (E530). Nevertheless, due to concerns regarding genotoxicity, the European Food Safety Authority (EFSA) has recently prohibited the utilization of titanium dioxide (E171) as a food additive (EFSA et al., 2021a). However, the FDA has recently classified zinc o□ide, iron oxide, and copper oxide as nutritional dietary supplements for animal feeds, and they are now considered to be generally recognized as safe (GRAS). In addition to the aforementioned group, copper oxide and di-copper oxide should be included. These substances have been approved by the EFSA Panel on Additives and Products or Substances used in Animal Feed (FEEDAP), which has determined that they do not present any health risks to consumers (He et al., 2019). The minimal regulation of nano food is primarily attributed to the intricate nature of nanomaterials and the case-by-case legislative procedures (SCENIHR, 2012a; b; He et al., 2019). In addition, the variation in levels of toxicity, effects, and risks associated with different categories of nanomaterials is another factor that contributes to the lack of knowledge (Deng et al., 2018).

The FDA in the United States and the SCENIHR (Scientific Committee on Emerging and Newly Identified Health Risks) in the European Union are the primary regulatory and legislative bodies responsible for overseeing food nanotechnology. The regulations of the European Union emphasize that food ingredients containing nanomaterials must undergo a safety assessment to ensure their safety for human consumption (Tinkle et al., 2014). In addition, nanofood and food ingredients are comprehensively regulated by the European Union Novel Foods Regulation (EC 258-97) and the REACH (Registration, Evaluation, Authorization and Restriction of Chemicals) framework, which serves as the primary regulatory instrument for the European Union. The FDA is also involved in the legislation of nanotechnology in the food industry, albeit to a lesser extent. The European Union (EU) is at the forefront of implementing norms and regulations for the control of nanomaterials, particularly in relation to their use as food additives. The EFSA Scientific Network of Risk Assessment of Nanotechnologies in Food and Feed, which is marking its 10th anniversary in 2020, aims to develop new guidelines for regulating and assessing the risks associated with nanomaterials in agricultural, food, and feed products (EFSA, 2021a). However, it is necessary for China and Japan, who are key players in nanotechnology, to establish regulations regarding the utilization and proper disposal of nanomaterials (Brien & Cummins, 2010; Nile et al., 2020).

However, in 2017, the Taiwan FDA implemented new regulations regarding the safety assessment and pre-market approval of nanomaterials used in food packaging (Chemical Watch, 2017).

In addition, the Organization for Economic Co-operation and Development (OECD) established the "Program on Manufactured Nanomaterials" and the "Test Guidelines Program" in 2006. These programs aim to create a network and encourage collaboration among different disciplines to advance the development and use of nanomaterials in manufacturing. In order to advance this objective, the OECD initiated a new three-year project (NANOMET) in 2020, which is being financed by the European Union. In addition, the GHS Subcommittee of the United Nations (UN) working group is considering the application of the GHS criteria to synthetic nanomaterials. The UN Strategic Approach to International Chemicals Management (SAICM) serves as the global platform for addressing nano-safety concerns. The SAICM introduced a resolution specifically focused on nanotechnology and included new initiatives related to nanomaterials in its Global Plan of Action (Karlaganis et al., 2019).

Various geographic regions, such as the EU, Switzerland, Thailand, and USA, are enacting nano-regulations in accordance with the laws of the World Trade Organization (WTO). Maintaining the main regulatory objective of safeguarding consumers and the environment is of utmost importance. Furthermore, the legislation should not be used as a protectionist measure to favor domestic industries (Karlaganis et al., 2019).

Additional research is required to establish guidelines for the safe use and disposal of nanomaterials. This research should prioritize the development of methods for identifying and characterizing nanomaterials, as well as assessing their long-term risks to both humans and the environment. The stakeholders must undertake these tasks in order to make progress on the various issues. In addition, public engagement will be crucial to guarantee a transparent and productive discussion (Chena & Yadab, 2011).

The fisheries and aquaculture sectors stand to gain significant advantages, and current toxicological data indicate that nanomaterials are comparatively less dangerous than other chemicals already employed in these industries. Moreover, the anticipated risk to workers using nanotechnology products in fisheries and aquaculture appears to be within acceptable thresholds. These workers are anticipated to have a higher probability of being exposed to nanomaterials found in finished products, as opposed to raw materials that contain loose particles. Regarding exposure to nanomaterials, the occupational health and safety risks for most workers in these industries appear to be low or comparable to those of the general public. This is because nanomaterials are currently being used in various commercial products (Handy, 2012, pp. 1-29). The World Health Organization (WHO) released a set of guidelines on safeguarding workers from potential hazards associated with manufactured nanomaterials. These guidelines consist of a recommended roster of occupational exposure limits (WHO, 2017).

The governance of nanotechnology applications in aquaculture, as well as other agricultural sectors, should be examined in relation to intellectual property and technology transfer models, as well as the need for consistent financial investment in research and development. Furthermore, it is imperative to increase efforts in order to facilitate the integration and dissemination of technology among developing nations. Nanotechnology inherently requires the participation and collaboration of multiple disciplines, including researchers, industry, government, and the general public (Chena & Yadab, 2011).

In the past ten years, countries such as India, Brazil, and South Africa have made significant investments in the research and development of nanotechnology applications in various agricultural sectors and food systems (Gruere et al., 2011). Public-private sector partnerships and collaborations between developed and developing countries can be advantageous in achieving the primary objectives of the aquaculture sector, leading to mutual and global benefits.

6.9. CONCLUSIONS

At present, a range of nanotechnological applications have been utilized to enhance the aquaculture industry, potentially contributing significantly to its future development and sustainability. Currently, there exist numerous prospective uses for nanomaterials in the fisheries and aquaculture sector. Promising areas in this field include fish health management, incorporating nanoscale ingredients, using nanotechnology in aquaculture feeds and food packaging, as well as developing value-added products, stress reduction techniques, and health management applications. Presently, the majority of these applications are in a nascent phase, and the primary inhibiting factor for their widespread adoption is the exorbitant cost.

In order to achieve sustainable development of nanotechnology in the fisheries and aquaculture industries, it is necessary to conduct a thorough evaluation of its potential adverse effects. This entails conducting a meticulous examination of the life cycle and shelf life of new nanomaterials, as well as assessing the potential health and environmental risks associated with their exposure, release, and deposition. Nevertheless, these concerns should not deter us from endeavoring to incorporate nanotechnological applications in the aquaculture industry, provided that they are accompanied by meticulous surveillance and regulated usage. This approach will facilitate endeavors to mitigate risks and optimize advantages.

REFERENCES

1. Abarca, R. L., Rodriguez, F. J., Guarda, A., Galotto, M. J., & Bruna, J. E. (2016). Characterization of beta-cyclodextrin inclusion complexes containing an essential oil component. *Food Chemistry, 196*, 968–975. https://doi.org/10.1016/j.foodchem.2015.10.023.
2. Abd El-Naby, A. S., Al-Sagheer, A. A., Negm, S. S., & Naiel, M. A. E. (2020). Dietary combination of chitosan nanoparticle and thymol affects feed utilization, digestive enzymes, antioxidant status, and intestinal morphology of *Oreochromis niloticus*. *Aquaculture, 515*, 734577. https://doi.org/10.1016/j.aquaculture.2019.734577.
3. Abd El-Naby, F. S., Naiel, M. A. E., Al-Sagheer, A. A., & Negm, S. S. (2019). Dietary chitosan nanoparticles enhance the growth, production performance, and immunity in *Oreochromis niloticus*. *Aquaculture, 501*, 82–89. https://doi.org/10.1016/j.aquaculture.2018.11.014.
4. Abdolahpur, M. F., Chupani, L., Vijver, M. G., Vancova, M., & Peijnenburg, W. J. (2019). Analytical approaches for characterizing and quantifying engineered nanoparticles in biological matrices from an (eco) toxicological perspective: Old challenges, new methods and techniques. *The Science of the Total Environment, 660*, 1283–1293. https://doi.org/10.1016/j.scitotenv.2019.01.105.
5. Acevedo-Fani, A., Salvia-Trujillo, L., Rojas-Grau, M. A., & Martín-Belloso, O. (2015). Edible films from essential-oil-loaded nanoemulsions: Physicochemical characterization and antimicrobial properties. *Food Hydrocolloids, 47*, 168–177. https://doi.org/10.1016/j.foodhyd.2015.01.032.
6. Alboofetileh, M., Rezaei, M., Hosseini, H., & Abdollahi, M. (2016). Efficacy of activated alginate-based nanocomposite films to control *Listeria monocytogenes* and spoilage flora in rainbow trout slice. *Journal of Food Science & Technology, 53*(1), 521–530. https://doi.org/10.1007/s13197-015-2015-9.
7. Alishahi, A., Mirvaghefi, A., Tehrani, M. R., Farahmand, H., Koshio, S., Dorkoosh, F. A., & Elsabee, M. Z. (2011). Chitosan nanoparticle to carry vitamin C through the gastrointestinal tract and induce the non-specific immunity system of rainbow trout (*Oncorhynchus mykiss*). *Carbohydrate Polymers, 86*(1), 142–146. https://doi.org/10.1016/j.carbpol.2011.04.028.
8. Anderson, A., Allan, S., Petersen, A., & Wilkinson, C. (2005). The framing of nanotechnologies in the British newspaper press. *Science Communication, 27*(2), 200–220. https://doi.org/10.1177/1075547005281472.
9. ANSES (French Agency for Food, Environmental and Occupational Health

& Safety), Anastasi, E., Riviere, G., & Teste, B. (2019). Nanomaterials in food-prioritization & assessment. *EFSA Journal, 7*(S2), Article e170909. https://doi.org/10.2903/j.efsa.2019.e170909.

10. Aouini, F., Trombini, C., Volland, V., Elcafsi, M., & Blasco, J. (2018). Assessing lead toxicity in the clam *Ruditapes philippinarum*: Bioaccumulation and biochemical responses. *Ecotoxicology and Environmental Safety, 158*, 193–203. https://doi.org/10.1016/j.ecoenv.2018.04.033.
11. Asaikkutti, A., Bhavan, P. S., Vimala, K., Karthik, M., & Cheruparam-Bath, P. (2016). Dietary supplementation of green synthesized manganese-oxide nanoparticles and its effect on growth performance, muscle composition and digestive enzyme activities of the giant freshwater prawn *Macrobrachium rosenbergii*. *Journal of Trace Elements in Medicine & Biology, 35*, 7–17. https://doi.org/10.1016/j.jtemb.2016.01.005.
12. Asharani, P. V., Wu, Y. L., Gong, Z., & Valiyaveettil, S. (2008). Toxicity of silver nanoparticles in zebrafish models. *Nanotechnology, 19*(25), 255102. https://doi.org/10.1088/0957-4484/19/25/255102.
13. Ashouri, S., Keyvanshokooh, S., Salati, A. P., Johari, S. A., & Pasha-Zanoosi, H. (2015). Effects of different levels of dietary selenium nanoparticles on growth performance, muscle composition, blood biochemical profiles and antioxidant status of common carp (*Cyprinus carpio*). *Aquaculture, 446*(1), 25–29. https://doi.org/10.1016/j.aquaculture.2015.04.021.
14. Ashraf, M., Aklakur, M., Sharma, R., Ahmad, A., & Khan, M. (2011). Nanotechnology as a novel tool in fisheries and aquaculture development: A review. *Iranica Journal of Energy and Environment, 2*(3), 258–261. https://doi.org/10.5829/idosi.ijee.2011.02.03.2272.
15. Atchudan, R., Edison, T. N. J. I., Perumal, S., Karthikeyan, D., & Lee, Y. R. (2017). Effective photocatalytic degradation of anthropogenic dyes using graphene oxide grafting titanium dioxide nanoparticles under UV-light irradiation. *Journal of Photochemistry and Photobiology A: Chemistry, 333*, 92–104. https://doi.org/10.1016/j.jphotochem.2016.10.021.
16. Azizi-Lalabadi, M., Hashemi, H., Feng, J., & Jafari, S. M. (2020). Carbon nanomaterials against pathogens; the antimicrobial activity of carbon nanotubes, graphene/graphene oxide, fullerenes, and their nanocomposites. *Advances in Colloid and Interface Science, 284*, 102250. https://doi.org/10.1016/j.cis.2020.102250.
17. Baalousha, M., Cornelis, G., Kuhlbusch, T. A. J., Lynch, I., Nickel, C., Peijnenburg, W., & Van Den Brink, N. W. (2016). Modeling nanomaterial fate and uptake in the environment: Current knowledge and future trends.

Environmental Science: Nano, 3(2), 323–345. https://doi.org/10.1039/C5EN00207A.

18. Baldissera, M. D., Souza, C. F., da Silva, A. S., Velho, M. C., Ourique, A. F., & Baldisserotto, B. (2020a). Benefits of nanotechnology: Dietary supplementation with nerolidol-loaded nanospheres increases survival rates, reduces bacterial loads, and prevents oxidative damage in brains of Nile tilapia experimentally infected by *Streptococcus agalactiae*. *Microbial Pathogenesis, 141*, 103989. https://doi.org/10.1016/j.micpath.2020.103989.
19. Baldissera, M. D., Souza, C. F., Velho, M. C., Bassotto, V. A., Ourique, A. F., Da Silva, A. S., & Baldisserotto, B. (2020b). Nanospheres as a technological alternative to suppress hepatic cellular damage and impaired bioenergetics caused by nerolidol in Nile tilapia (*Oreochromis niloticus*). *Naunyn-Schmiedeberg's Archives of Pharmacology, 393*(5), 751–759. https://doi.org/10.1007/s00210-020-01824-2.
20. Baranwal, A., Srivastava, A., Kumar, P., Bajpai, V. K., Maurya, P. K., & Chandra, P. (2018). Prospects of nanostructure materials and their composites as antimicrobial agents. *Frontiers in Microbiology, 9*, 422. https://doi.org/10.3389/fmicb.2018.00422.
21. Behera, T., Nanda, P. K., Mohanty, C., Mohapatra, D., Swain, P., Das, B. K., Routray, P., Mishra, B. K., & Sahoo, S. K. (2010). Parenteral immunization of fish, *Labeo rohit* with Poly D, L-lactide-co-glycolic acid (PLGA) encapsulated antigen microparticles promotes innate and adaptive immune responses. *Fish & Shellfish Immunology, 28*(2), 320–325. https://doi.org/10.1016/j.fsi.2009.11.009.
22. Bello, D., & Leong, D. T. (2017). A decade of nanotoxicology: Assessing the impact on human health and the environment. *NanoImpact, 7*, 15–16. https://doi.org/10.1016/j.impact.2017.04.001.
23. Berntssen, M. H. G., Julshamn, K., & Lundebye, A. K. (2010). Chemical contaminants in aquafeeds and Atlantic salmon (*Salmo salar*) following the use of traditional–versus alternative feed ingredients. *Chemosphere, 78*(6), 637–646. https://doi.org/10.1016/j.chemosphere.2009.12.021.
24. Bhattacharyya, A., Reddy, J., Hasan, M. M., Adeyemi, M. M., & Marye, R. R. (2015). Nanotechnology—A unique future technology in aquaculture for the food security. *International Journal of Bioassays, 4*(7), 4115–4126.
25. Binks, B. P. (2007). Colloidal particles at liquid interfaces. *Physical Chemistry Chemical Physics, 9*, 6298–6299. Retrieved from: https://dce.uni-sofia.bg/files/publications/2007/pdf/pdf/2007-09-MB-NC-KD-PK-Electrocapillary-Attraction.pdf (accessed on 3 August 2024).

26. Borges, O., Silva, M., de Sousa, A., Borchard, G., Junginger, H. E., & Cordeiro da Silva, A. (2008). Alginate coated chitosan nanoparticles are an effective subcutaneous adjuvant for hepatitis B surface antigen. *International Immunopharmacology, 8*(13–14), 1773–1780. https://doi.org/10.1016/j.intimp.2008.08.013.
27. Borgogna, M., Bellich, B., & Cesaro, A. (2011). Marine polysaccharides in microencapsulation and application to aquaculture: "From sea to sea". *Marine Drugs, 9*(12), 2572–2604. https://doi.org/10.3390/md9122572.
28. Boxall, A., Chaudhry, Q., Sinclair, C., Jones, A., Aitken, R., Jefferson, B., & Watts, C. (2007). Current and future predicted environmental exposure to manufactured nanoparticles. *Report by the Central Science Laboratory (CSL) York for the Department of the Environment and Rural Affairs (DEFRA), U.K.* Retrieved from: http://randd.defra.gov.uk/Document.asp?Document=CB01098_6270_FRP.pdf (accessed on 3 August 2024).
29. Brien, N. O., & Cummins, E. (2010). Ranking initial environmental and human health risk resulting from environmentally relevant nanomaterials. *Journal of Environmental Science & Health Part A, 45*(8), 992–1007. https://doi.org/10.1080/10934521003772410.
30. Bundschuh, M., Filser, J., Lüderwald, S., McKee, M. S., Metreveli, G., Schaumann, G. E., Schulz, R., & Wagner, S. (2018). Nanoparticles in the environment: Where do we come from, where do we go to? *Environmental Sciences Europe, 30*(1), 1–17. https://doi.org/10.1186/s12302-018-0132-6.
31. Canesi, L., & Corsi, I. (2016). Effects of nanomaterials on marine invertebrates. *The Science of the Total Environment, 565*, 933–940. https://doi.org/10.1016/j.scitotenv.2016.01.085.
32. Canesi, L., Balbi, T., Fabbri, R., Damonte, G., Salis, A., Volland, M., & Blasco, J. (2017). Biomolecular coronas in invertebrate species: Implications in the environmental impact of nanoparticles. *NanoImpact, 8*, 89–98. https://doi.org/10.1016/j.impact.2017.08.001.
33. Canesi, L., Ciacci, C., & Balbi, T. (2015). Interactive effects of nanoparticles with other contaminants in aquatic organisms: Friend or foe? *Marine Environmental Research, 111*, 128–134. https://doi.org/10.1016/j.marenvres.2015.03.010.
34. Cerrillo, C., Barandika, G., Igartua, A., Areitioaurtena, O., & Mendoza, G. (2017). Key challenges for nanotechnology: Standardization of ecotoxicity testing. *Journal of Environmental Science & Health–Part C, 35*(2), 104–126. https://doi.org/10.1080/10590501.2017.1298361.
35. Chandra, P. (2016). *Nano-Biosensors for Personalized and Onsite*

Biomedical Diagnosis (1st ed., p. 637). The Institution of Engineering and Technology.

36. Chaudhry, Q., Watkins, R., & Castle, L. (2017). Nanotechnologies in food: What, why and how? In Q. Chaudhry, L. Castle, & R. Watkins (Eds.), *Nanotechnologies in Food* (pp. 1–19).
37. Chellaram, C., Murugaboopathi, G., John, A. A., Sivakumar, R., Ganesan, S., Krithika, S., & Priya, G. (2014). Significance of nanotechnology in food industry. *APCBEE Procedia, 8*, 109–113. https://doi.org/10.1016/j.apcbee.2014.03.010.
38. Chemical Watch. (2017). Taiwan FDA issues guidelines on nanomaterials in food packaging: Pre-market approval requirements updated. Retrieved from: https://chemicalwatch.com/54578/taiwan-fda-issues-guidelines-on-nanomaterials-in-food-packaging (accessed on 3 August 2024).
39. Chen, B., Zou, L., Wu, Z., & Sun, M. (2016). The application of quantum dots in aquaculture pollution detection. *Toxicological and Environmental Chemistry, 98*(3–4), 385–394. 0
40. Chen, H., Weiss, J., & Shahidi, F. (2006). Nanotechnology in nutraceuticals and functional foods. *Food Technology, 60*(3), 30–36.
41. Chena, H., & Yadab, R. (2011). Nanotechnologies in agriculture: New tools for sustainable development. *Trends in Food Science & Technology, 22*(11), 585–594. https://doi.org/10.1016/j.tifs.2011.09.004.
42. Cheng, A. C., Chen, Y. Y., & Chen, J. C. (2008). Dietary administration of sodium alginate and κ-carrageen/an enhances the innate immune response of brown-marbled grouper *Epinephelus fuscoguttatus* and its resistance against *Vibrio alginolyticus*. *Veterinary Immunology and Immunopathology, 121*(3–4), 206–215. https://doi.org/10.1016/j.vetimm.2007.09.011.
43. Cheng, T. C., Yao, K. S., Yeh, N., Chang, C. I., Hsu, H. C., Chien, Y. T., & Chang, C. Y. (2009). Visible light activated bactericidal effect of TiO2/Fe3O4 magnetic particles on fish pathogens. *Surface and Coatings Technology, 204*(6–7), 1141–1144. https://doi.org/10.1016/j.surfcoat.2009.06.050.
44. Chifiriuc, M. C., Kamerzan, C., & Lazar, V. (2017). Essential oils and nanoparticles: New strategy to prevent microbial biofilms. In A. Ficai & A. M. Grumezescu (Eds.), *Nanostructures for Antimicrobial Therapy: A Volume in Micro and Nano Technologies* (pp. 279–291). Elsevier.
45. Chiu, S. T., Tsai, R. T., Hsu, J. P., Liu, C. H., & Cheng, W. (2008). Dietary sodium alginate administration to enhance the non-specific immune responses, and disease resistance of the juvenile grouper *Epinephelus*

fuscoguttatus. *Aquaculture, 277*(1–2), 66–72. https://doi.org/10.1016/j.aquaculture.2008.01.032.

46. Christofoli, M., Costa, E. C. C., Bicalho, K. U., de Cassia-Domingues, V., Peixoto, M. F., Fernandes-Alves, C. C., Araujo, W. L., & de Melo-Cazal, C. (2015). Insecticidal effect of nanoencapsulated essential oils from *Zanthoxylum rhoifolium* (Rutaceae) in *Bemisia tabaci* populations. *Industrial Crops and Products, 70*, 301–308. 0
47. Ciobanu, A., Mallard, I., Landy, D., Brabie, G., Nistor, D., & Fourmentin, S. (2012). Inclusion interactions of cyclodextrins and cross-linked cyclodextrin polymers with linalool and camphor in *Lavandula angustifolia* essential oil. *Carbohydrate Polymers, 87*(3), 1963–1970. https://doi.org/10.1016/j.carbpol.2011.10.005.
48. Cortes-Rojas, D. F., Souza, C. R. F., & Oliveira, W. P. (2014). Encapsulation of eugenol-rich clove extract in solid lipid carriers. *Journal of Food Engineering, 127*, 34–42. https://doi.org/10.1016/j.jfoodeng.2013.11.027.
49. Cossu, A., Wang, M. S., Chaudhari, A., & Nitin, N. (2015). Antifungal activity against *Candida albicans* of starch Pickering emulsion with thymol or amphotericin B in suspension and calcium alginate films. *International Journal of Pharmaceutics, 493*(1–2), 233–242. https://doi.org/10.1016/j.ijpharm.2015.07.065.
50. Cui, J., Yu, B., Zhao, Y., Zhu, W., Li, H., Lou, H., & Zhai, G. (2009). Enhancement of oral absorption of curcumin by self-microemulsifying drug delivery systems. *International Journal of Pharmaceutics, 371*(1–2), 148–155. https://doi.org/10.1016/j.ijpharm.2008.12.009.
51. da Rosa, C. G., de Oliveira Brisola Maciel, M. V., de Carvalho, S. M., de Melo, A. P. Z., Jummes, B., da Silva, T., Martelli, S. M., Viletti, M. A., Bertoldi, F. C., & Manique Barreto, P. L. (2015). Characterization and evaluation of physicochemical and antimicrobial properties of zein nanoparticles loaded with phenolic monoterpenes. *Colloids and Surfaces A: Physicochemical and Engineering Aspects, 481*, 337–344. https://doi.org/10.1016/j.colsurfa.2015.05.019.
52. Daniel, S. K., Sureshkumar, V., & Sivakumar, M. (2016). Nano ice based on silver nanoparticles for fish preservation. *International Journal of Fisheries and Aquaculture Studies, 4*(5), 162–167.
53. Das, R. K., Kasoju, N., & Bora, U. (2010). Encapsulation of curcumin in alginate-chitosan-pluronic composite nanoparticles for delivery to cancer cells. *Nanomedicine: Nanotechnology, Biology and Medicine, 6*(1), 153–160. https://doi.org/10.1016/j.nano.2009.05.009.
54. De Oliveira, E. F., Paula, H. C. B., & de Paula, R. C. M. (2014). Alginate/

cashew gum nanoparticles for essential oil encapsulation. *Colloids and Surfaces B, 113*, 146–151. https://doi.org/10.1016/j.colsurfb.2013.08.038.

55. Deng, H., Zhang, Y., & Yu, H. (2018). Nanoparticles considered as mixtures for toxicological research. *Journal of Environmental Science and Health–Part C Environmental Carcinogenesis & Ecotoxicology Reviews, 36*(1), 1–20. https://doi.org/10.1080/10590501.2018.1418792.
56. Dickinson, E. (2010). Food emulsions and foams: Stabilization by particles. *Current Opinion in Colloid & Interface Science, 15*(1–2), 40–49. https://doi.org/10.1016/j.cocis.2009.11.001.
57. Donbrow, M. (1991). *Microcapsules and Nanoparticles in Medicine and Pharmacy*. CRC Press.
58. Dursun, S., Erkan, N., & Ye ilta , M. (2010). Application of natural biopolymer-based nanocomposite films in seafood. *Journal of Fisheries Science, 4*(1), 50–77.
59. EFSA FAF (Panel EFSA Panel on Food Additives and Flavourings), Younes, M., Aquilina, G., Castle, L., et al. (2021b). Scientific Opinion on the safety assessment of titanium dioxide (E171) as a food additive. *EFSA Journal, 19*(5), 6585. https://doi.org/10.2903/j.efsa.2021.6585.
60. ElBasuini, M. F., El-Hais, A. M., Dawood, M. A. O., Abou-Zeid, A. E. S., ElDamrawy, S. Z., Khalafalla, M. M. E. S., Khalafalla, S., Ishikawa, K. M., & Dossou, I. S. (2017). Effects of dietary copper nanoparticles and vitamin C supplementations on growth performance, immune response, and stress resistance of red sea bream, *Pagrus major*. *Aquaculture Nutrition, 23*(6), 1329–1340. https://doi.org/10.1111/anu.12508.
61. Eldridge, J. H., Hammond, C. J., Meulbroek, J. A., Staas, J. K., Gilley, R. M., & Tice, T. R. (1990). Controlled vaccine release in gut-associated lymphoid tissues. I. Orally administered biodegradable microspheres target the Peyer's patches. *Journal of Controlled Release, 11*(1–3), 205–214. https://doi.org/10.1016/0168-3659(90)90133-E.
62. Elieh-Ali-Komi, D., & Hamblin, M. R. (2016). Chitin and chitosan: Production and application of versatile biomedical nanomaterials. *International Journal of Advanced Research, 4*(3), 411–427.
63. Erdem, B., Dayangac, A., Kıray, E., & Duygu, D. (2018). Biosynthesis of silver nanoparticles from *Aeromonas sobria* and antibacterial activity against fish pathogens. *International Journal of Environmental Science and Technology, 16*(9), 1–6. https://doi.org/10.1007/s13762-018-1944-z.
64. FAO. (2018). *The State of World Fisheries and Aquaculture 2018–Meeting the Sustainable Development Goals*. Rome: Food & Agriculture Organization. Retrieved from: http://www.fao.org/3/i9540en/i9540en.pdf

(accessed on 3 August 2024).

65. Ferosekhan, S., Gupta, S., Singh, A., Rather, M., Kumari, R., Kothari, D., Pal, K. A., & Jadhao, S. B. (2014). RNA-loaded chitosan nanoparticles for enhanced growth immunostimulation and disease resistance in fish. *Current Nanoscience, 10*(3), 453–464. https://doi.org/10.2174/15734137 10666140115220300.
66. Florence, A. T., Hillery, A. M., Hussain, N., & Jani, P. U. (1995). Nanoparticles as carriers for oral peptide absorption: Studies on particle uptake and fate. *Journal of Controlled Release, 36*(1–2), 39–46. https://doi.org/10.1016/0168-3659(95)00059-H.
67. Foster, H. A., Ditta, I. B., Varghese, S., & Steele, A. (2011). Photocatalytic disinfection using titanium dioxide: Spectrum and mechanism of antimicrobial activity. *Applied Microbiology and Biotechnology, 90*(6), 1847–1868. https://doi.org/10.1007/s00253-011-3213-7.
68. Fujiki, K., Matsuyama, H., & Yano, T. (1994). Protective effect of sodium alginate against bacterial infection in common carp, *Cyprinus carpio* L. *Journal of Fish Diseases, 17*(4), 349–355. https://doi.org/10.1111/j.1365-2761.1994.tb00230.x.
69. Garcia-Negrete, C. A., Blasco, J., Volland, M., Rojas, T. C., Hampel, M., Lapresta Fernández, A., Jimenez-de Haro, M. C., Soto, M., & Fernandez, A. (2013). Behavior of Au-citrate nanoparticles in seawater and accumulation in bivalves at environmentally relevant concentrations. *Environmental Pollution, 174*, 134–141. https://doi.org/10.1016/j.envpol.2012.11.014.
70. Gaysinsky, S., Davidson, P. M., Bruce, B. D., & Weiss, J. (2005). Growth inhibition of *Escherichia coli* O157: H7 and *Listeria monocytogenes* by carvacrol and eugenol encapsulated in surfactant micelles. *Journal of Food Protection, 68*(12), 2559–2566. https://doi.org/10.4315/0362-028X-68.12.2559.
71. Giese, B., Klaessig, F., Park, B., Kaegi, R., Steinfeldt, M., Wigger, H., von Gleich, A., & Gottschalk, F. (2018). Risks, release and concentrations of engineered nanomaterials in the environment. *Scientific Reports, 8*, 1565. https://doi.org/10.1038/s41598-018-19275-4.
72. Gottschalk, F., Sonderer, T., Scholz, R. W., & Nowack, B. (2009). Modelled environmental concentrations of manufactured nanomaterials (TiO_2, ZnO, Ag, CNT, fullerenes) for different regions. *Environmental Science & Technology, 43*(24), 9216–9222. https://doi.org/10.1021/es9015553.
73. Griffitt, R. J., Luo, J., Gao, J., Bonzongo, J. C., & Barber, D. S. (2008). Effects of particle composition and species on toxicity of metallic nanomaterials in aquatic organisms. *Environmental Toxicology &*

Chemistry, 27(9), 1972–1978. https://doi.org/10.1897/08-002.1.

74. Gruere, G., Narrod, C., & Abbott, L. (2011). *Agriculture, Food, and Water Nanotechnologies for the Poor: Opportunities and Constraints*. International Food Policy Research Institute (IFPRI).
75. Halimi, M., Alishahi, M., Abbaspour, M. R., Ghorbanpoor, M., & Tabandeh, M. R. (2019). Valuable method for production of oral vaccine by using alginate and chitosan against *Lactococcus garvieae/Streptococcus iniae* in rainbow trout (*Oncorhynchus mykiss*). *Fish & Shellfish Immunology, 90*, 431–439. https://doi.org/10.1016/j.fsi.2019.05.020.
76. Handy, R. D. (2012). FSBI briefing paper: Nanotechnology in fisheries and aquaculture. Fisheries Society of the British Isles.
77. Handy, R. D., & Poxton, M. G. (1993). Nitrogen pollution in mariculture–to□icity and e□cretion of nitrogenous compounds by marine fish. *Reviews in Fish Biology and Fisheries, 3*(3), 205–241. https://doi.org/10.1007/BF00043929.
78. Handy, R. D., & Shaw, B. J. (2007). Toxic effects of nanoparticles and nanomaterials: Implications for public health, risk assessment and the public perception of nanotechnology. *Health, Risk & Society, 9*(2), 125–144. https://doi.org/10.1080/13698570701306807.
79. Handy, R. D., Al-Bairuty, G., Al-Jubory, A., & Henry, T. B. (2011). Effects of manufactured nanomaterials on fishes: A target organ and body systems physiology approach. *Journal of Fish Biology, 79*(4), 821–853. https://doi.org/10.1111/j.1095-8649.2011.03080.x.
80. Handy, R., von der Kammer, F., Lead, J., Hassellov, M., Owen, R., & Crane, M. (2008). The ecotoxicology and chemistry of manufactured nanoparticles. *Ecotoxicology, 17*, 287–314. https://doi.org/10.1007/s10646-008-0199-8.
81. He, X., Deng, H., & Hwang, H. (2019). The current application of nanotechnology in food and agriculture. *Journal of Food and Drug Analysis, 27*(1), 1–21. https://doi.org/10.1016/j.jfda.2018.12.002.
82. Hill, L. E., Gomes, C., & Taylor, T. M. (2013). Characterization of beta-cyclodextrin inclusion complexes containing essential oils (trans-cinnamaldehyde, eugenol, cinnamon bark, and clove bud extracts) for antimicrobial delivery applications. *LWT–Food Science and Technology, 51*(1), 86–93. https://doi.org/10.1016/j.lwt.2012.11.011.
83. Hosseini, S. F., Rezaei, M., Zandi, M., & Farahmandghavi, F. (2015). Fabrication of bio nanocomposite films based on fish gelatin reinforced with chitosan nanoparticles. *Food Hydrocolloids, 44*, 172–182. https://doi.org/10.1016/j.foodhyd.2014.09.004.

84. Hosseini, S. F., Rezaei, M., Zandi, M., & Farahmandghavi, F. (2016). Development of bioactive fish gelatin/chitosan nanoparticles composite films with antimicrobial properties. *Food Chemistry, 194*, 1266–1274. https://doi.org/10.1016/j.foodchem.2015.09.004.
85. Hu, C., Lu, T., Chen, F., & Zhang, R. (2013). A brief review of graphene-metal oxide composites synthesis and applications in photocatalysis. *Journal of Chinese Advanced Materials and Society, 1*(1), 21–39. https://doi.org/10.1080/22243682.2013.771917.
86. Huang, S., Wang, L., Liu, L., Hou, Y., & Li, L. (2015). Nanotechnology in agriculture, livestock, and aquaculture in China: A review. *Agronomy for Sustainable Development, 35*(2), 369–400. https://doi.org/10.1007/s13593-014-0274-x.
87. Huang, Y., Mei, L., Chen, X., & Wang, Q. (2018). Recent developments in food packaging based on nanomaterials. *Nanomaterials, 8*(10), 830. https://doi.org/10.3390/nano8100830.
88. Huttenhuis, H. B. T., Ribeiro, A. S. P., Bowden, T. J., Van Bavel, C., Taverne-Thiele, A. J., & Rombout, J. H. W. M. (2006). The effect of oral immuno-stimulation in juvenile carp (*Cyprinus carpio* L.). *Fish & Shellfish Immunology, 21*(3), 261–271. https://doi.org/10.1016/j.fsi.2005.12.002.
89. Irie, T., Watarai, S., Iwasaki, T., & Kodama, H. (2005). Protection against experimental *Aeromonas salmonicida* infection in carp by oral immunization with bacterial antigen entrapped liposomes. *Fish & Shellfish Immunology, 18*(3), 235–242. https://doi.org/10.1016/j.fsi.2004.07.006.
90. Ismael, N. E. M., Abd El-hameed, S. A. A., Salama, A. M., Naiel, M. A. E., & Abdel-Latif, H. M. R. (2021). The effects of dietary clinoptilolite and chitosan nanoparticles on growth, body composition, haemato-biochemical parameters, immune responses, and antioxidative status of Nile tilapia exposed to imidacloprid. *Environmental Science & Pollution Research, 28*, 29535–29550. https://doi.org/10.1007/s11356-021-12693-4.
91. Izquierdo, M. S., Ghrab, W., Roo, J., Hamre, K., Hernandez-Cruz, C. M., Bernardini, G., Terova, G., & Saleh, R. (2017). Organic, inorganic and nanoparticles of Se, Zn and Mn in early weaning diets for gilthead seabream (*Sparus aurata*; Linnaeus, 1758). *Aquaculture Research, 48*(6), 2852–2867. https://doi.org/10.1111/are.13119.
92. Jain, A., Ranjan, S., Dasgupta, N., & Ramalingam, C. (2018). Nanomaterials in food and agriculture: An overview on their safety concerns and regulatory issues. *Critical Reviews in Food Science and Nutrition, 58*(2), 297–317. https://doi.org/10.1080/10408398.2016.1160363.

93. Jani, P., Halbert, G. W., Langridge, J., & Florence, A. T. (1990). Nanoparticles uptake by rat gastrointestinal mucosa: Quantitation and particles size dependency. *Journal of Pharmaceutics & Pharmacology, 42*(12), 821–826. https://doi.org/10.1111/j.2042-7158.1990.tb07033.x.
94. Jeong, S. H., Hwang, Y. H., & Yi, S. C. (2005). Antibacterial properties of padded PP/PE nonwovens incorporating nano-sized silver colloids. *Journal of Materials Science, 40*(20), 5413–5418. https://doi.org/10.1007/s10853-005-4340-2.
95. Jiang, X., Valdeperez, D., Nazarenus, M., Wang, Z., Stellacci, F., Parak, W. J., & del Pino, P. (2015). Future perspectives towards the use of nanomaterials for smart food packaging and quality control. *Particle & Particle Systems Characterization, 32*(4), 408–416. https://doi.org/10.1002/ppsc.201400192.
96. Jimenez-Fernandez, E., Ruyra, A., Roher, N., Infante, C., & Fernandez-Diaz, C. (2014). Nanoparticles as a novel delivery system for vitamin C administration in aquaculture. *Aquaculture, 432*, 426–433. https://doi.org/10.1016/j.aquaculture.2014.03.006.
97. Jimenez, A., Sanchez-Gonzalez, L., Desobry, S., Chiralt, A., & Tehrany, E. A. (2014). Influence of nanoliposomes incorporation on properties of film forming dispersions and films based on corn starch and sodium caseinate. *Food Hydrocolloids, 35*, 159–169. https://doi.org/10.1016/j.foodhyd.2013.05.006.
98. Joukar, F., Hosseini, S. M. H., Moosavi-Nasab, M., Mesbahi, G. R., & Behzadnia, A. (2017). Effect of Farsi gum-based antimicrobial adhesive coatings on the refrigeration shelf life of rainbow trout fillets. *LWT–Food Science and Technology, 80*, 1–9. https://doi.org/10.1016/j.lwt.2017.01.074.
99. Jovanovic, B., Anastasova, L., Rowe, E. W., Zhang, Y., Clapp, A. R., & Palic, D. (2011). Effects of nanosized titanium dioxide on innate immune system of fathead minnow (*Pimephales promelas* Rafinesque, 1820). *Ecotoxicology and Environmental Safety, 74*(4), 675–683. https://doi.org/10.1016/j.ecoenv.2010.10.017.
100. Ju-Nam, Y., & Lead, J. R. (2008). Manufactured nanoparticles: An overview of their chemistry, interactions and potential environmental implications. *The Science of the Total Environment, 400*(1–3), 396–414. https://doi.org/10.1016/j.scitotenv.2008.06.042.
101. Kaphle, A., Navya, P. N., Umapathi, A., & Daima, H. K. (2018). Nanomaterials for agriculture, food, and environment: Applications, toxicity, and regulation. *Environmental Chemistry Letters, 16*(1), 43–58.

https://doi.org/10.1007/s10311-017-0662-y.

102. Karlaganis, G., Liechti, R., Teparkum, S., Aungkavattana, P., & Indaraprasirt, R. (2019). Nanoregulation along the product life cycle in the EU, Switzerland, Thailand, the USA, and intergovernmental organizations, and its compatibility with WTO law.
103. Kepiro, I. E., Marzuoli, I., Hammond, K., Ba, X., Lewis, H., Shaw, M., Gunnoo, S. B., De Santis, E., Kapinska, U., Pagliara, S., Holmes, M. A., Lorenz, C. D., Hoogenboom, B. W., Fraternali, F., & Ryadnov, M. G. (2020). Engineering chirally blind protein pseudocapsids into antibacterial persisters. *ACS Nano, 14*(2), 1609–1622. https://doi.org/10.1021/acsnano.9b06814.
104. Khosravi-Katuli, K., Prato, E., Lofrano, G., Guida, M., Vale, G., & Libralato, G. (2017). Effects of nanoparticles in species of aquaculture interest. *Environmental Science & Pollution Research, 24*(21), 17326–17346. https://doi.org/10.1007/s11356-017-9360-3.
105. Kim, J., Park, N., Na, J. H., & Han, J. (2016). Development of natural insect-repellent loaded halloysite nanotubes and their application to food packaging to prevent *Plodia interpunctella* infestation. *Journal of Food Science, 81*(8), E1956–E1965. https://doi.org/10.1111/1750-3841.13373.
106. Kitiyodom, S., Yata, T., Yostawornkul, J., Kaewmalun, S., Nittayasut, N., Suktham, K., Surassmo, S., Namdee, K., Rodkhum, C., & Pirarat, N. (2019). Enhanced efficacy of immersion vaccination in tilapia against columnaris disease by chitosan-coated "pathogen-like" mucoadhesive nanovaccines. *Fish & Shellfish Immunology, 95*, 213–219. https://doi.org/10.1016/j.fsi.2019.09.064.
107. Klaine, S. J., Alvarez, P. J. J., Batley, G. E., Fernandes, T. F., Handy, R. D., Lyon, D. Y., Mahendra, S., McLaughlin, M. J., & Lead, J. R. (2008). Nanomaterials in the environment: Behavior, fate, bioavailability, and effects. *Environmental Toxicology & Chemistry, 27*(9), 1825–1851. https://doi.org/10.1897/08-090.1.
108. Kole, S., Qadiri, S. S. N., Shin, S. M., Kim, W. S., Lee, J., & Jung, S. J. (2019). Nanoencapsulation of inactivated-viral vaccine using chitosan nanoparticles: Evaluation of its protective efficacy and immune modulatory effects in olive flounder (*Paralichthys olivaceus*) against viral haemorrhagic septicaemia virus (VHSV) infection. *Fish & Shellfish Immunology, 91*, 136–147. https://doi.org/10.1016/j.fsi.2019.05.017.
109. Korni, F. M. M., & Khalil, F. (2017). Effect of ginger and its nanoparticles on growth performance, cognition capability, immunity, and prevention of

motile *Aeromonas septicaemia* in *Cyprinus carpio* fingerlings. *Aquaculture Nutrition, 23*(6), 1492–1499. https://doi.org/10.1111/anu.12526.

110. Kumar, H., Kuča, K., Bhatia, S. K., Saini, K., Kaushal, A., Verma, R., Bhalla, T. C., & Kumar, D. (2020). Applications of nanotechnology in sensor-based detection of foodborne pathogens. *Sensors, 20*(7), 1966. https://doi.org/10.3390/s20071966.

111. Kumar, S. R., Ahmed, V. I., Parameswaran, V., Sudhakaran, R., Babu, V. S., & Hameed, A. S. (2008). Potential use of chitosan nanoparticles for oral delivery of DNA vaccine in Asian sea bass (*Lates calcarifer*) to protect from *Vibrio (Listonella) anguillarum*. *Fish & Shellfish Immunology, 25*(1–2), 47–56. https://doi.org/10.1016/j.fsi.2007.12.004.

112. Kunjiappan, S., Bhattacharjee, C., & Chowdhury, R. (2015). Hepatoprotective and antioxidant effects of *Azolla microphylla* based gold nanoparticles against acetaminophen induced toxicity in a freshwater common carp fish (*Cyprinus carpio* L.). *Nanomedicine Journal, 2*(2), 88–110.

113. Kushnerova, N. F., Fomenko, S. E., Sprygin, V. J., Kushnerova, T. V., Khotimchenko, Y. S., Kondrateva, E. V., & Drugova, L. A. (2010). An extract from the brown alga *Laminaria japonica*: A promising stress-protective preparation. *Russian Journal of Marine Biology, 36*(3), 209–214. https://doi.org/10.1134/S1063074010030077.

114. Kuswandi, B. (2016). Nanotechnology in food packaging. In S. Ranjan, N. Dasgupta, & E. Lichtfouse (Eds.), *Nanoscience in Food and Agriculture 1: Sustainable Agriculture Reviews* (Vol. 20). Cham: Springer. https://doi.org/10.1007/978-3-319-39303-2_6.

115. Lai, F., Wissing, S. A., Muller, R. H., & Fadda, A. M. (2006). *Artemisia arborescens* L essential oil-loaded solid lipid nanoparticles for potential agricultural application: Preparation and characterization. *The AAPS Journal, 7*(1), E10–E18. https://doi.org/10.1208/pt070102.

116. Leclercq, L., & Nardello-Rataj, V. (2016). Pickering emulsions based on cyclodextrins: A smart solution for antifungal azole derivatives topical delivery. *European Journal of Pharmaceutical Sciences, 82*, 126–137. https://doi.org/10.1016/j.ejps.2015.11.017.

117. Lee, M. H., & Park, H. J. (2015). Preparation of halloysite nanotubes coated with Eudragit for a controlled release of thyme essential oil. *Journal of Applied Polymer Science, 132*(46). https://doi.org/10.1002/app.42771.

118. Lee, S. B., Kim, J. Y., Kim, K., Ahn, K. J., Kim, T., & Oh, J. M. (2020). Encapsulation and release control of fish pathogen utilizing cross-linked alginate networks and clay nanoparticles for use with a potential oral

vaccination. *Applied Sciences, 10*(8), 2679. https://doi.org/10.3390/app10082679.

119. Leon-Rodríguez, L., Luzardo-Alvarez, A., Blanco-Mendez, J., Lamas, J., & Leiro, J. (2013). Biodegradable microparticles covalently linked to surface antigens of the scuticociliate parasite *P. dicentrarchi* promote innate immune responses in vitro. *Fish & Shellfish Immunology, 34*(1), 236–243. https://doi.org/10.1016/j.fsi.2012.10.029.
120. Letchford, K., Liggins, R., & Burt, H. (2008). Solubilization of hydrophobic drugs by methoxy poly (ethylene glycol)-block-polycaprolactone diblock copolymer micelles: Theoretical and experimental data and correlations. *Journal of Pharmaceutical Sciences, 97*(3), 1179–1190. https://doi.org/10.1002/jps.21037.
121. Li, J., Xu, X., Chen, Z., Wang, T., Lu, Z., Hu, W., & Li, W. (2018). Zein/gum Arabic nanoparticle-stabilized Pickering emulsion with thymol as an antibacterial delivery system. *Carbohydrate Polymers, 200*, 416–426. https://doi.org/10.1016/j.carbpol.2018.08.025.
122. Li, L., Lin, S. L., Deng, L., & Liu, Z. G. (2013). Potential use of chitosan nanoparticles for oral delivery of DNA vaccine in black seabream *Acanthopagrus schlegelii* Bleeker to protect from *Vibrio parahaemolyticus*. *Journal of Fish Diseases, 36*(12), 987–995. https://doi.org/10.1111/jfd.12032.
123. Li, Q., Wu, P., & Shang, J. K. (2014). Nanostructured visible-light photocatalysts for water purification. In A. Street, R. Sustich, J. Duncan, & N. Savage (Eds.), *Nanotechnology Applications for Clean Water: Solutions for Improving Water Quality* (pp. 297–317). Norwich, NY: William Andrew. https://doi.org/10.1016/B978-1-4557-3116-9.00019-6.
124. Liakos, I. L., Grumezescu, A. M., Holban, A. M., Florin, I., D'Autilia, F., Carzino, R., Bianchini, P., & Athanassiou, A. (2016). Polylactic acid-lemongrass essential oil nanocapsules with antimicrobial properties. *Pharmaceuticals, 9*(3), 42. https://doi.org/10.3390/ph9030042.
125. Liu, F., & Tang, C. H. (2016). Reprint of Soy glycinin as food-grade Pickering stabilizers: Part. III. Fabrication of gel-like emulsions and their potential as sustained-release delivery systems for β-carotene. *Food Hydrocolloids, 60*, 631–640. https://doi.org/10.1016/j.foodhyd.2016.05.004.
126. Liu, L., Cui, Z., Ma, Q., Cui, W., & Zhang, X. (2016). One-step synthesis of magnetic iron-aluminum oxide/graphene oxide nanoparticles as a selective adsorbent for fluoride removal from aqueous solution. *RSC Advances, 6*(13), 10783–10791.
127. Majumder, D., & Dash, G. (2017). Application of nanotechnology in

fisheries and aquaculture. *Everyman Science, 51*(6), 358–364. Retrieved from: http://www.sciencecongress.nic.in/pdf/e-book/Feb17-March17.pdf (accessed on 3 August 2024).

128. Malheiros, D. F., Sarquis, I. R., Ferreira, I. M., Mathews, P. D., Mertins, O., & Tavares Dias, M. (2020). Nanoemulsions with oleoresin of *Copaifera reticulata* (Leguminosae) improve anthelmintic efficacy in the control of monogenean parasites when compared to oleoresin without nanoformulation. *Journal of Fish Diseases, 43*, 687–695. https://doi.org/10.1111/jfd.13168.

129. Marku, D., Wahlgren, M., Rayner, M., Sjoo, M., & Timgren, A. (2012). Characterization of starch Pickering emulsions for potential applications in topical formulations. *The Internet Journal of Pharmacology, 428*, 1–7. https://doi.org/10.1016/j.ijpharm.2012.01.031.

130. Matos, M., Laca, A., Rea, F., Iglesias, O., Rayner, M., & Gutierrez, G. (2018). O/W emulsions stabilized by OSA-modified starch granules versus non-ionic surfactant: Stability, rheological behavior and resveratrol encapsulation. *Journal of Food Engineering, 222*, 207–217. https://doi.org/10.1016/j.jfoodeng.2017.11.009.

131. Matos, M., Marefati, A., Gutierrez, G., Wahlgren, M., & Rayner, M. (2016). Comparative emulsifying properties of octenyl succinic anhydride (OSA)-modified starch: Granular form vs dissolved state. *PLoS One, 11*, Article e0160140. https://doi.org/10.1371/journal.pone.0160140.

132. Matranga, V., & Corsi, I. (2012). Toxic effects of engineered nanoparticles in the marine environment: Model organisms and molecular approaches. *Marine Environmental Research, 76*, 32–40. https://doi.org/10.1016/j.marenvres.2012.01.006.

133. McClements, D. J. (2011). Edible nano emulsions: Fabrication, properties, and functional performance. *Soft Matter, 7*(6), 2297–2316. https://doi.org/10.1039/C0SM00549E.

134. McClements, D. J. (2017). The future of food colloids: Next-generation nanoparticle delivery systems. *Current Opinion in Colloid & Interface Science, 28*, 7–14. https://doi.org/10.1016/j.cocis.2016.12.002.

135. McClements, D. J., & Rao, J. (2011). Food-grade nanoemulsions: Formulation, fabrication, properties, performance, biological fate, and potential toxicity. *Critical Reviews in Food Science and Nutrition, 51*(4), 285–330. https://doi.org/10.1080/10408398.2011.559558.

136. Meghani, N., Dave, S., & Kumar, A. (2020). Introduction to nanofood. In U. Hebbar, S. Ranjan, N. Dasgupta, & R. Kumar Mishra (Eds.), *Nano-*

Food Engineering (Food Engineering Series). Cham: Springer. https://doi.org/10.1007/978-3-030-44552-2_11.

137. Metreveli, G., Frombold, B., Seitz, F., Grün, A., Phillippe, A., Rosenfeldt, R. R., Bundschuh, M., Schulz, R., Manz, W., & Schaumann, G. E. (2016). Impact of chemical composition of ecotoxicological test media on the stability and aggregation status of silver nanoparticles. *Environmental Science: Nano, 3*, 418–433. https://doi.org/10.1039/C5EN00152H.
138. Mihindukulasuriya, S. D. F., & Lim, L. T. (2014). Nanotechnology development in food packaging: A review. *Trends in Food Science & Technology, 40*(2), 149–167. https://doi.org/10.1016/j.tifs.2014.09.009.
139. Mitchell, J., & Trivedi, V. (2010). Pharmaceutical nanomaterials: The preparation of solid core drug delivery systems (SCDDS). *Journal of Pharmaceutics & Pharmacology, 62*, 1457–1458. https://doi.org/10.1111/j.2042-7158.2010.01178.x.
140. Moghimipour, E., Ramezani, Z., & Handali, S. (2013). Solid lipid nanoparticles as a delivery system for *Zataria multiflora* essential oil: Formulation and characterization. *Current Drug Delivery, 10*, 151–157. https://doi.org/10.2174/156720181131002000.
141. Mohammadi, N., & Tukmechi, A. (2015). The effects of iron nanoparticles in combination with *Lactobacillus casei* on growth parameters and probiotic counts in rainbow trout (*Oncorhynchus mykiss*) intestine. *Journal of Veterinary Research, 70*, 47–53. https://doi.org/10.22059/JVR.2015.52968.
142. Mohan, C. O., Ravishankar, C. N., Lalitha, K. V., & Gopal, T. S. (2012). Effect of chitosan edible coating on the quality of double filleted Indian oil sardine (*Sardinella longiceps*) during chilled storage. *Food Hydrocolloids, 26*(1), 167–174. https://doi.org/10.1016/j.foodhyd.2011.05.005.
143. Mohanraj, V. J., & Chen, Y. (2006). Nanoparticles—A review. *Tropical Journal of Pharmaceutical Research, 5*(1), 561–573. https://doi.org/10.4314/tjpr.v5i1.14634.
144. Mongillo, J. F. (2007). *Nanotechnology 101*. Westport, CT: Greenwood Press.
145. Moore, M. N. (2006). Do nanoparticles present ecotoxicological risk for the health of the aquatic environment? *Environment International, 32*, 967–976. https://doi.org/10.1016/j.envint.2006.06.014.
146. Morris, K. (2009). Nanotechnology crucial in fighting infectious disease. *The Lancet Infectious Diseases, 9*, 215. https://doi.org/10.1016/S1473-3099(09)70100-8.

147. Motamedi, E., Talebi-Atouei, M., & Kassaee, M. Z. (2014). Comparison of nitrate removal from water via graphene oxide coated Fe, Ni, and Co nanoparticles. *Materials Research Bulletin, 54*, 34–40. https://doi.org/10.1016/j.materresbull.2014.02.019.

148. Mukherjee, A., Maity, A., Pramanik, P., Shubha, K., Joshi, D. C., & Wani, S. H. (2019). Public perception about use of nanotechnology in agriculture. In M. Ghorbanpour & S. H. Wani (Eds.), *Advances in Phytonanotechnology* (pp. 405–418). Academic Press.

149. Muralisankar, T., Saravana-Bhavan, P., Radhakrishnan, S., Seenivasan, C., & Srinivasan, V. (2016). The effect of copper nanoparticles supplementation on freshwater prawn *Macrobrachium rosenbergii* post larvae. *Journal of Trace Elements in Medicine & Biology, 34*, 39–49. https://doi.org/10.1016/j.jtemb.2015.12.003.

150. Naguib, M., Mahmoud, U. M., Mekkawy, I. A., & Sayed, A. E. D. H. (2020). Hepatotoxic effects of silver nanoparticles on *Clarias gariepinus*: Biochemical, histopathological and histochemical studies. *Toxicology Reports, 7*, 133–141. https://doi.org/10.1016/j.toxrep.2020.01.002.

151. Naiel, M. A. E., Ismael, N. E. M., Abd El-hameed, S. A. A., & Amer, M. S. (2020). The antioxidative and immunity roles of chitosan nanoparticle and vitamin C supplemented diets against imidacloprid toxicity on *Oreochromis niloticus*. *Aquaculture, 523*, 735219. https://doi.org/10.1016/j.aquaculture.2020.735219.

152. Nangmenyi, G., & Economy, J. (2014). Nanometallic particles for oligodynamic microbial disinfection. In A. Street, R. Sustich, J. Duncan, & N. Savage (Eds.), *Nanotechnology Applications for Clean Water: Solutions for Improving Water Quality* (pp. 283–295). Norwich, NY: William Andrew.

153. Ngwabebhoh, F. A., Erdagi, S. I., & Yildiz, U. (2018). Pickering emulsions stabilized nanocellulosic-based nanoparticles for coumarin and curcumin nanoencapsulations: In vitro release, anticancer and antimicrobial activities. *Carbohydrate Polymers, 201*, 317–328. https://doi.org/10.1016/j.carbpol.2018.08.079.

154. Nickols-Richardson, S. M., & Piehowski, K. E. (2008). Nanotechnology in nutritional sciences. *Minerva Biotechnologica, 20*(3), 117.

155. Nile, S. H., Baskar, V., Selvaraj, D., Nile, A., Xiao, J., & Kai, G. (2020). Nanotechnologies in food science: Applications, recent trends, and future perspectives. *Nano-Micro Letters, 12*, 45. https://doi.org/10.1007/s40820-020-0383-9.

156. Nowack, B., & Bucheli, T. D. (2007). Occurrence, behavior and effects

of nanoparticles in the environment. *Environmental Pollution, 150*, 5–22. https://doi.org/10.1016/j.envpol.2007.06.006.

157. O'Callaghan, K. A., & Kerry, J. P. (2016). Preparation of low- and medium-molecular weight chitosan nanoparticles and their antimicrobial evaluation against a panel of microorganisms, including cheese-derived cultures. *Food Control, 69*, 256–261. https://doi.org/10.1016/j.foodcont.2016.05.005.
158. Ogunkalu, O. A. (2019). Utilization of nanotechnology in aquaculture and seafood sectors. *EJFST, 19*, 26–33.
159. Otoni, C. G., Avena-Bustillos, R. J., Olsen, C. W., Bilbao-Sainz, C., & McHugh, T. H. (2016). Mechanical and water barrier properties of isolated soy protein composite edible films as affected by carvacrol and cinnamaldehyde micro and nanoemulsions. *Food Hydrocolloids, 57*, 72–79. https://doi.org/10.1016/j.foodhyd.2016.01.012.
160. Otoni, C. G., de Moura, M. R., Aouada, F. A., Camilloto, G. P., Cruz, R. S., Lorevice, M. V., Soares, F. F., & Mattoso, L. H. (2014). Antimicrobial and physical-mechanical properties of pectin/papaya puree/cinnamaldehyde nanoemulsion edible composite films. *Food Hydrocolloids, 41*, 188–194. https://doi.org/10.1016/j.foodhyd.2014.04.013.
161. Owen, R., & Handy, R. D. (2007). Formulating the problems for environmental risk assessment of nanomaterials. *Environmental Science & Technology, 41*, 5582–5588. https://doi.org/10.1021/es072598h.
162. Ozogul, Y., Yuvka, I., Ucar, Y., Durmus, M., Kosker, A. R., Oz, M., & Ozogul, F. (2017). Evaluation of effects of nanoemulsion based on herb essential oils (rosemary, laurel, thyme, and sage) on sensory, chemical, and microbiological quality of rainbow trout (*Oncorhynchus mykiss*) fillets during ice storage. *LWT-Food Science and Technology, 75*, 677–684. https://doi.org/10.1016/j.lwt.2016.10.009.
163. Parris, N., Cooke, P. H., & Hicks, K. B. (2005). Encapsulation of essential oils in zein nanospherical particles. *Journal of Agricultural and Food Chemistry, 53*, 4788–4792. https://doi.org/10.1021/jf040492p.
164. Pathakoti, K., Manubolu, M., & Hwang, H. M. (2017). Nanostructures: Current uses and future applications in food science. *Journal of Food and Drug Analysis, 25*(2), 245–253. https://doi.org/10.1016/j.jfda.2017.02.004.
165. Patolsky, F., Zheng, G. F., Hayden, O., Lakadamyali, M., Zhuang, X. W., & Lieber, C. M. (2004). Electrical detection of single viruses. *Proceedings of the National Academy of Sciences, 101*, 14017–14022. https://doi.org/10.1073/pnas.0406159101.
166. Pavela, R., Maggi, F., Ngahang Kamte, S. L., Rakotosaona, R., Rasoanaivo,

P., Nicoletti, M., Canale, A., & Benelli, G. (2017). Chemical composition of *Cinnamosma madagascariensis* (Cannelaceae) essential oil and its larvicidal potential against the filariasis vector *Culex quinquefasciatus* Say. *South African Journal of Botany, 108*, 359–363. https://doi.org/10.1016/j.sajb.2016.08.017.

167. Peters, S. E., & Brain, C. H. (2009). Benefits of a soy lecithin-based nanotechnology for animals and humans. In Q. Huang, P. Given, & M. Qian (Eds.), *Micro/Nanoencapsulation of Active Food Ingredients* (pp. 183–197). American Chemical Society. https://doi.org/10.1021/bk-2009-1007.ch012.

168. Rajeshkumar, S., Venkatesan, C., Sarathi, M., Sarathbabu, V., Thomas, J., Anver Basha, K., & Hameed, A. S. S. (2009). Oral delivery of DNA construct using chitosan nanoparticles to protect the shrimp from white spot syndrome virus (WSSV). *Fish & Shellfish Immunology, 26*, 429–437. https://doi.org/10.1016/j.fsi.2009.01.003.

169. Ramsden, C. S., Smith, T. J., Shaw, B. S., & Handy, R. D. (2009). Dietary exposure to titanium dioxide nanoparticles in rainbow trout (*Oncorhynchus mykiss*): No effect on growth, but subtle biochemical disturbances in the brain. *Ecotoxicology, 18*, 939–951. https://doi.org/10.1007/s10646-009-0357-7.

170. Ramsden, J. J. (2018). Nanotechnology in a modern economy. In J. J. Ramsden (Ed.), *Applied Nanotechnology: The Conversion of Research Results to Products* (pp. 61–80). Oxford: William Andrew Applied Science Publishers.

171. Rather, M. A., Sharma, R., Aklakur, M., Ahmad, S., Kumar, N., Khan, M., & Ramya, V. L. (2011). Nanotechnology: A novel tool for aquaculture and fisheries development. A prospective mini-review. *Fish & Aquac. J., 16*.

172. Reilly, R. M. (2007). Carbon nanotubes: Potential benefits and risks of nanotechnology in nuclear medicine. *Journal of Nuclear Medicine, 48*, 1039. https://doi.org/10.2967/jnumed.107.041723.

173. Reis, C. P., Neufeld, R. J., & Veiga, F. (2017). Preparation of drug-loaded polymeric nanoparticles. In L. P. Balogh (Ed.), *Nanomedicine in Cancer* (pp. 197–240). Singapore: Pan Stanford Publishing Pte Ltd.

174. Rhim, J. W., Park, H. M., & Ha, C. S. (2013). Bio-nanocomposites for food packaging applications. *Progress in Polymer Science, 38*, 1629–1652. https://doi.org/10.1016/j.progpolymsci.2013.05.008.

175. Roy, S., Kumar, V., Barman, D., & Mandal, S. (2012). Nanotechnology and its application in fisheries and aquaculture. *AQUA International, 20*(6), 2–3.

176. Sayed, A. E. D. H., Mekkawy, I. A., Mahmoud, U. M., & Nagiub, M. (2020). Histopathological and histochemical effects of silver nanoparticles on the gills and muscles of African catfish (*Clarias gariepinus*). *Scientific African, 7*, Article e00230. https://doi.org/10.1016/j.sciaf.2019.e00230.

177. SCENIHR. (2012a). *Nanomaterials: Case by Case Safety Approach for Breakthrough Technology* (IP-12–1050_EN). European Commission, Brussels. Retrieved from: https://ec.europa.eu/commission/presscorner/detail/en/IP_12_1050 (accessed on 3 August 2024).

178. SCENIHR. (2012b). *Nanomaterials: Commission Proposes Case by Case Approach to Assessment* (MEMO-12-732_EN). European Commission, Brussels. Retrieved from: https://ec.europa.eu/commission/presscorner/detail/en/MEMO_12_732 (accessed on 3 August 2024).

179. SCENIHR. (2012c). *Second Regulatory Review on Nanomaterials* (COM (2012) 572 final). European Commission. Retrieved from: https://eur-lex.europa.eu/LexUriServ/LexUriServ.do?uri=COM:2012:0572:FIN:EN:PDF (accessed on 3 August 2024).

180. Sekhon, B. S. (2014). Nanotechnology in agri-food production: An overview. *Nanotechnology, Science and Applications, 7*, 31. https://doi.org/10.2147/NSA.S39406.

181. Selck, H., Handy, R. D., Fernandes, T. F., Klaine, S. J., & Petersen, E. J. (2016). Nanomaterials in the aquatic environment: A European Union–United States perspective on the status of ecotoxicity testing, research priorities, and challenges ahead. *Environmental Toxicology & Chemistry, 35*, 1055–1067. https://doi.org/10.1002/etc.3385.

182. Semalty, A., Semalty, M., Rawat, M. S. M., & Franceschi, F. (2010). Supramolecular phospholipids-polyphenolics interactions: The PHYTOSOME strategy to improve the bioavailability of phytochemicals. *Fitoterapia, 81*, 306–314. https://doi.org/10.1016/j.fitote.2009.11.001.

183. Sendra, M., Blasco, J., & Araujo, C. (2018a). Is the cell wall of marine phytoplankton a protective barrier or a nanoparticle interaction site? Toxicological responses of *Chlorella autotrophica* and *Dunaliella salina* to Ag and CeO_2 nanoparticles. *Ecological Indicators, 95*, 1053–1067. https://doi.org/10.1016/j.ecolind.2017.08.050.

184. Sendra, M., Este, M. P., Garrido, I. M., Gatica, J. M., & Blasco, J. (2017a). CeO_2 NPs, toxic or protective to phytoplankton? Charge of nanoparticles and cell wall as factors which cause changes in cell complexity. *The Science of the Total Environment, 590–591*, 304–315. https://doi.org/10.1016/j.scitotenv.2017.03.007.

185. Sendra, M., Este, M. P., Gatica, J. M., Garrido, I. M., & Blasco, J.

(2017b). Direct and indirect effects of silver nanoparticles on freshwater and marine microalgae (*Chlamydomonas reinhardtii* and *Phaeodactylum tricornutum*). *Chemosphere, 179*, 279–289. https://doi.org/10.1016/j.chemosphere.2017.03.123.

186. Sendra, M., Volland, M., Balbi, T., Fabbri, R., Yeste, M. P., Gatica, J. M., Canesi, L., & Blasco, J. (2018b). Cytotoxicity of CeO_2 nanoparticles using in vitro assay with *Mytilus galloprovincialis* hemocytes: Relevance of zeta potential, shape and biocorona formation. *Aquatic Toxicology, 200*, 13–20. https://doi.org/10.1016/j.aquatox.2018.04.011.
187. Sendra, M., Yeste, M. P., Garrido, I. M., Gatica, J. M., & Blasco, J. (2017c). Homoagglomeration and heteroagglomeration of TiO_2, in nanoparticle and bulk form, onto freshwater and marine microalgae. *The Science of the Total Environment, 592*, 403–411. https://doi.org/10.1016/j.scitotenv.2017.03.127.
188. Shah, B. R., & Mraz, J. (2019). Advances in nanotechnology for sustainable aquaculture and fisheries. *Reviews in Aquaculture*, *12*(2), 925–942. https://doi.org/10.1111/raq.12356.
189. Shah, B. R., Li, Y., Jin, W., An, Y., He, L., Li, Z., Xu, W., & Li, B. (2016). Preparation and optimization of Pickering emulsion stabilized by chitosan-tripolyphosphate nanoparticles for curcumin encapsulation. *Food Hydrocolloids, 52*, 369–377. https://doi.org/10.1016/j.foodhyd.2015.07.015.
190. Shah, C., Mishra, B., Kumar, M., Priyadarsini, K., & Bajaj, P. (2008). Binding studies of curcumin to polyvinyl alcohol/polyvinyl alcohol hydrogel and its delivery to liposomes. *Current Science, 95*(10), 1426–1432.
191. Shao, P., Zhang, H., Niu, B., & Jin, W. (2018). Physical stabilities of taro starch nanoparticles stabilized Pickering emulsions and the potential application of encapsulated tea polyphenols. *International Journal of Biological Macromolecules, 118*, 2032–2039. https://doi.org/10.1016/j.ijbiomac.2018.07.076.
192. Sharif, R. M., Haghighi, M., & Bazari Moghaddam, S. (2017). Study on nanoparticles of Aloe vera extract on growth performance, survival rate and body composition in Siberian sturgeon (*Acipenser baerii*). *Iranian Journal of Fisheries Sciences, 16*, 457–468.
193. Shaw, B., & Handy, R. D. (2011). Physiological effects of nanoparticles on fish: A comparison of nanometals versus metal ions. *Environment International, 37*(6), 1083–1097. https://doi.org/10.1016/j.envint.2011.03.009.

194. Sibaja-Luis, A. I., Ramos-Campos, E. V., de Oliveira, J. L., & Fernandes, L. (2019). Trend in aquaculture sciences: From now to use of nanotechnology for disease control. *Reviews in Aquaculture, 11*(1), 119–132. https://doi.org/10.1111/raq.12229.

195. Siegrist, M., Stampfli, N., Kastenholz, H., & Keller, C. (2008). Perceived risks and perceived benefits of different nanotechnology foods and nanotechnology food packaging. *Appetite, 51*(2), 283–290. https://doi.org/10.1016/j.appet.2008.02.020.

196. Siqueira-Lima, P. S., Araujo, A. A. S., Lucchese, A. M., Quintans, J. S., Menezes, P. P., Alves, P. B., de Lucca, W., Santos, M. R. V., Bonjardim, L. R., & Quintans-Junior, L. J. (2014). β-cyclode□trin comple□containing *Lippia grata* leaf essential oil reduces orofacial nociception in mice—Evidence of possible involvement of descending inhibitory pain modulation pathway. *Basic and Clinical Pharmacology and Toxicology, 114*, 188–196. https://doi.org/10.1111/bcpt.12145.

197. Skjermo, J., & Bergh, Ø. (2004). High-M alginate immunostimulation of Atlantic halibut (*Hippoglossus hippoglossus* L.) larvae using Artemia for delivery, increases resistance against vibriosis. *Aquaculture, 238*, 107–113. https://doi.org/10.1016/j.aquaculture.2004.05.038.

198. Skjolding, L. M., Sørensen, S. N., Hartmann, N. B., Hjorth, R., Hansen, S. F., & Baun, A. (2016). A critical review of aquatic ecotoxicity testing of nanoparticles—The quest for disclosing nanoparticle effects. *Angewandte Chemie, 55*, 15224–15239. https://doi.org/10.1002/anie.201604964.

199. Solans, C., Izquierdo, P., Nolla, J., Azemar, N., & Garcia-Celma, M. J. (2005). Nanoemulsions. *Current Opinion in Colloid & Interface Science, 10*(3–4), 102–110. https://doi.org/10.1016/j.cocis.2005.06.004.

200. Sotelo-Boyas, M. E., Correa-Pacheco, Z. N., Bautista-Banos, S., & Corona-Rangel, M. L. (2017b). Physicochemical characterization of chitosan nanoparticles and nanocapsules incorporated with lime essential oil and their antibacterial activity against food-borne pathogens. *Lebensmittel-Wissenschaft & Technologie, 77*, 15–20. https://doi.org/10.1016/j.lwt.2016.11.022.

201. Sotelo-Boyas, M., Correa-Pacheco, Z., Bautista-Banos, S., & Gomez, Y. (2017a). Release study and inhibitory activity of thyme essential oil-loaded chitosan nanoparticle and nanocapsules against foodborne bacteria. *International Journal of Biological Macromolecules, 103*, 409–414. https://doi.org/10.1016/j.ijbiomac.2017.05.063.

202. Souza, C. F., Baldissera, M. D., Santos, R. C., Raffin, R. P., & Baldisserotto, B. (2017). Nanotechnology improves the therapeutic efficacy of *Melaleuca*

alternifolia essential oil in experimentally infected *Rhamdia quelen* with *Pseudomonas aeruginosa*. *Aquaculture, 473*, 169–171. https://doi.org/10.1016/j.aquaculture.2017.02.014.

203. Stone, V., Nowack, B., Baun, A., van den Brink, N., von der Kammer, F., Dusinska, M., Handy, R., Hankin, S., Hassellov, M., Joner, E., & Fernandes, T. F. (2010). Nanomaterials for environmental studies: Classification, reference material issues and strategies for physico-chemical characterisation. *The Science of the Total Environment, 408*, 1745–1754. https://doi.org/10.1016/j.scitotenv.2009.10.035.

204. Stromme, M., Brohede, U., Atluri, R., & Garcia-Bennett, A. E. (2009). Mesoporous silica-based nanomaterials for drug delivery: Evaluation of structural properties associated with release rate. *Wiley Interdisciplinary Reviews: Nanomedicine and Nanobiotechnology, 1*, 140–148. https://doi.org/10.1002/wnan.13.

205. Sun, T. Y., Bornhoft, N. A., Hungerbuhler, K., & Nowack, B. (2016). Dynamic probabilistic modeling of environmental emissions of engineered nanomaterials. *Environmental Science & Technology, 50*(9), 4701–4711. https://doi.org/10.1021/acs.est.5b05828.

206. Suppan, S. (2011). Racing ahead: US agri-nanotechnology in the absence of regulation. Institute for Agriculture and Trade Policy. https://doi.org/10.17844/jphpi.v11i2.907.

207. Suptijah, P. Y., Gushagia, D., & Sukarsa, D. R. (2008). Study of inhibitory effects of chitosan on quality deterioration of catfish (*Pangasius hypopthalmus*) fillet at room temperature storage. *JPHPI, 11*(2), 89–101.

208. Swain, P., Sasmal, A., Nayak, S. K., Barik, S. K., Mishra, S. S., Mohapatra, K. D., Swain, S. K., Saha, J. N., Sen, A. K., & Jayasankar, P. (2016). Evaluation of selected metal nanoparticles on hatching and survival of larvae and fry of Indian major carp, rohu (*Labeo rohita*). *Aquaculture Research, 47*, 498–511. https://doi.org/10.1111/are.12510.

209. Swastawati, F., Wijayanti, I., & Susanto, E. (2008). Utilization of shrimp shell waste as edible coating to reduce environmental pollution. *Indonesian Journal of Environment Technology, 4*(4), 101–106.

210. Tafaghodi, M., Sajadi Tabasi, S. A., & Payan, M. (2007). Alginate microsphere as a delivery system and adjuvant for autoclaved *Leishmania major* and *Quillaja saponin*: Preparation and characterization. *Iranian Journal of Pharmaceutical Sciences, 3*, 61–68.

211. Takahashi, M., Uechi, S., Takara, K., Asikin, Y., & Wada, K. (2009). Evaluation of an oral carrier system in rats: Bioavailability and antioxidant properties of liposome encapsulated curcumin. *Journal of Agricultural*

and Food Chemistry, 57, 9141–9146. https://doi.org/10.1021/jf9013923.

212. Tandberg, J., Lagos, L., Ropstad, E., Smistad, G., Hiorth, M., & Winther-Larsen, H. C. (2018). The use of chitosan-coated membrane vesicles for immunization against salmonid rickettsial septicemia in an adult zebrafish model. *Zebrafish, 15*(4), 372–381. https://doi.org/10.1089/zeb.2017.1556.
213. Tapilatu, Y., Nugraheni, P. S., Ginzel, T., Latumahina, M., Limmon, G. V., & Budhijanto, W. (2016). Nano-chitosan utilization for fresh yellowfin tuna preservation. *Aquat. Procedia, 7*, 285–295. https://doi.org/10.1016/j.aqpro.2016.07.040.
214. Technavio. (2019). Global food nanotechnology market 2019–2023: Growing applications in nutraceuticals to boost the market. Toronto, Canada: Technavio. Retrieved from: https://www.technavio.com/report/global-food-nanotechnology-market-industry-analysis&utm_source=pressrelease&utm_medium=bw&utm_campaign=t12wk24&utm_content=IRTNTR30663 (accessed on 3 August 2024).
215. Thiruvengadam, M., Rajakumar, G., & Chung, I. M. (2018). Nanotechnology: Current uses and future applications in the food industry. *3 Biotech, 8*(1), 74. https://doi.org/10.1007/s13205-018-1104-7.
216. Tian, J., & Yu, J. (2011). Poly (lactic-co-glycolic acid) nanoparticles as candidate DNA vaccine carrier for oral immunization of Japanese flounder (*Paralichthys olivaceus*) against lymphocystis disease virus. *Fish & Shellfish Immunology, 30*, 109–117. https://doi.org/10.1016/j.fsi.2010.09.016.
217. Tinkle, S., Mcneil, S. E., Stefan, M., Bawa, R., & Borchard, G. (2014). Nanomedicines: Addressing the scientific and regulatory gap. *Annals of the New York Academy of Sciences, 1313*, 35–56. https://doi.org/10.1111/nyas.12403.
218. Torchilin, V. P. (2006). *Nanoparticulate as Drug Carriers*. London, UK: Imperial College Press.
219. Toyokawa, H., Nakao, A., Bailey, R. J., Nalesnik, M. A., Kaizu, T., Lemoine, J. L., Ikeda, A., Tomiyama, K., Papworth, G. D., Huang, L., Demetris, A. J., Starzl, T. E., & Murase, N. (2008). Relative contribution of direct and indirect allorecognition in developing tolerance after liver transplantation. *Liver Transplantation, 14*, 346–357. https://doi.org/10.1002/lt.21378.
220. Valencia-Sullca, C., Jimenez, M., Jimenez, A., Atares, L., Vargas, M., & Chiralt, A. (2016). Influence of liposome encapsulated essential oils on properties of chitosan films. *Polymer International, 65*(8), 979–987. https://doi.org/10.1002/pi.5143.

221. Vasile, C. (2018). Polymeric nanocomposites and nanocoating's for food packaging: A review. *Materials, 11*(10), 1834. https://doi.org/10.3390/ma11101834.

222. Vicari, T., Dagostim, A. C., Klingelfus, T., Limberger-Galvan, G., Sampaio-Monteiro, P., da Silva-Pereira, L., Silva-de Assis, H. C., & Cestari, M. M. (2018). Co-exposure to titanium dioxide nanoparticles ($NpTiO_2$) and lead at environmentally relevant concentrations in the Neotropical fish species *Hoplias intermedius*. *Toxicology Reports, 5*, 1032–1043. https://doi.org/10.1016/j.toxrep.2018.09.001.

223. Vignardi, C. P., Hasue, F. M., Sartorio, P. V., Cardoso, C. M., Machado, A. S., Passos, M. J., Santos, T. C. A., Nucci, J. M., Hewer, T. L. R., Watanabe, I. S., Gomes, V., & Phan, N. V. (2015). Genotoxicity, potential cytotoxicity and cell uptake of titanium dioxide nanoparticles in the marine fish *Trachinotus carolinus* (Linnaeus, 1766). *Aquatic Toxicology, 158*, 218–229. https://doi.org/10.1016/j.aquatox.2014.11.008.

224. Volland, M., Hampel, M., Katsumiti, A., Yeste, M. P., Gatica, J. M., Cajaraville, M., & Blasco, J. (2018). Synthesis methods influence characteristics, behavior and toxicity of bare CuO NPs compared to bulk CuO and ionic Cu after in vitro exposure of *Ruditapes philippinarum* hemocytes. *Aquatic Toxicology, 199*, 285–295. https://doi.org/10.1016/j.aquatox.2018.04.007.

225. Wacker, M. G., Proykova, A., & Santos, G. M. L. (2016). Dealing with nanosafety around the globe–regulation vs. innovation. *International Journal of Pharmaceutics, 509*, 95–106. https://doi.org/10.1016/j.ijpharm.2016.05.015.

226. Wang, Z., Leung, M. H. M., Kee, T. W., & English, D. S. (2009). The role of charge in the surfactant-assisted stabilization of the natural product curcumin. *Langmuir, 26*, 5520–5526. https://doi.org/10.1021/la903772e.

227. Winuprasith, T., Khomein, P., Mitbumrung, W., Suphantharika, M., Nitithamyong, A., & McClements, D. J. (2018). Encapsulation of vitamin D3 in Pickering emulsions stabilized by nano fibrillated mangosteen cellulose: Impact on in vitro digestion and bio accessibility. *Food Hydrocolloids, 83*, 153–164. https://doi.org/10.1016/j.foodhyd.2018.04.047.

228. World Health Organization (WHO). (2017). *WHO Guidelines on Protecting Workers from Potential Risks of Manufactured Nanomaterials*. Geneva, WHO. Retrieved from https://apps.who.int/iris/bitstream/handle/10665/259671/9789241550048-eng.pdf (accessed 3 August 2024).

229. Wu, J., & Ma, G. H. (2016). Recent studies of Pickering emulsions: Particles make the difference. *Small, 12*, 4633–4648. https://doi.

org/10.1002/smll.201600877.

230. Wu, J., Liu, H. G. S., Wang, S., Qin, Z., Chen, L., Zheng, Q., Liu, Q., & Zhang, Q. (2015b). The preparation, characterization, antimicrobial stability and in vitro release evaluation of fish gelatin films incorporated with cinnamon essential oil nanoliposomes. *Food Hydrocolloids, 43*, 427–435. https://doi.org/10.1016/j.foodhyd.2014.06.017.
231. Wu, L. P., Ficker, M., Christensen, J. B., Trohopoulos, P. N., & Moghimi, S. M. (2015a). Dendrimers in medicine: Therapeutic concepts and pharmaceutical challenges. *Bioconjugate Chemistry, 26*, 1198–1211. https://doi.org/10.1021/acs.bioconjchem.5b00031.
232. Wu, Y., Luo, Y., & Wang, Q. (2012). Antioxidant and antimicrobial properties of essential oils encapsulated in zein nanoparticles prepared by liquid-liquid dispersion method. *LWT–Food Science and Technology, 48*, 283–290. https://doi.org/10.1016/j.lwt.2012.03.027.
233. Xu, W., Huang, K., Jin, W., Luo, D., Liu, H., Li, Y., & Liu, X. (2018a). Catalytic and antibacterial properties of biosynthesized silver nanoparticles using native inulin. *RSC Advances, 8*, 28746–28752. https://doi.org/10.1039/C8RA03386B.
234. Xu, W., Jin, W., Huang, K., Huang, L., Lou, Y., Li, J., Liu, X., & Li, B. (2018b). Interfacial and emulsion stabilized behavior of lysozyme/xanthan gum nanoparticles. *International Journal of Biological Macromolecules, 117*, 280–286. https://doi.org/10.1016/j.ijbiomac.2018.05.187.
235. Yang, H. C., Wang, W. H., Huang, K. S., & Hon, M. H. (2010). Preparation and application of nano chitosan to finishing treatment with antimicrobial and anti-shrinking properties. *Carbohydrate Polymers, 79*(1), 176–179. https://doi.org/10.1016/j.carbpol.2009.07.045.
236. Yang, Z., Yue, G. H., & Wong, S. M. (2021). VNN disease and status of breeding for resistance to NNV in aquaculture. *Aquaculture Fisheries*. https://doi.org/10.1016/j.aaf.2021.04.001 (in press).
237. Yasumoto, S., Yoshimura, T., & Miyazaki, T. (2006). Oral immunization of common carp with a liposome vaccine containing *Aeromonas hydrophila* antigen. *Fish Pathology, 41*, 45–49. https://doi.org/10.3147/jsfp.41.45.
238. Ye, S., Jin, W., Huang, Q., Hu, Y., Shah, B. R., Li, Y., & Li, B. (2016). Development of Mag-FMBO in clay-reinforced KGM aerogels for arsenite removal. *International Journal of Biological Macromolecules, 87*, 77–84. https://doi.org/10.1016/j.ijbiomac.2016.01.087.
239. Yeh, S. P., Chang, C. A., Chang, C. Y., Liu, C. H., & Cheng, W. (2008). Dietary sodium alginate administration affects fingerling growth and resistance to *Streptococcus* sp. and iridovirus, and juvenile non-specific

immune responses of the orange-spotted grouper (*Epinephelus coioides*). *Fish & Shellfish Immunology, 25*, 19–27. https://doi.org/10.1016/j.fsi.2007.11.011.

240. Yi, T., Liu, C., Zhang, J., Wang, F., Wang, J., & Zhang, J. (2017). A new drug nanocrystal self-stabilized Pickering emulsion for oral delivery of silybin. *European Journal of Pharmaceutical Sciences, 96*, 420–427. https://doi.org/10.1016/j.ejps.2016.08.047.
241. Yun, S., Jun, J. W., Giri, S. S., Kim, H. J., Chi, C., Kim, S. G., Kim, S. W., & Park, S. C. (2017). Efficacy of PLGA microparticle-encapsulated formalin-killed *Aeromonas hydrophila* cells as a single-shot vaccine against *A. hydrophila* infection. *Vaccine, 35*(32), 3959–3965. https://doi.org/10.1016/j.vaccine.2017.06.005.
242. Zhang, Y., Niu, Y., Luo, Y., Ge, M., Yang, T., Yu, L., & Wang, Q. (2014). Fabrication, characterization and antimicrobial activities of thymol-loaded zein nanoparticles stabilized by sodium caseinate-chitosan hydrochloride double layers. *Food Chemistry, 142*, 269–275. https://doi.org/10.1016/j.foodchem.2013.07.058.
243. Zhou, X., Wang, Y., Gu, Q., & Li, W. (2009). Effects of different dietary selenium sources (selenium nanoparticle and selenomethionine) on growth performance, muscle composition and glutathione peroxidase enzyme activity of crucian carp (*Carassius auratus gibelio*). *Aquaculture, 291*, 78–81. https://doi.org/10.1016/j.aquaculture.2009.03.007.
244. Zhou, Y., Sun, S., Bei, W., Zahi, M. R., Yuan, Q., & Liang, H. (2018). Preparation and antimicrobial activity of oregano essential oil Pickering emulsion stabilized by cellulose nanocrystals. *International Journal of Biological Macromolecules, 112*, 7–13. https://doi.org/10.1016/j.ijbiomac.2018.01.102.

Chapter 7

Applications of Machine Learning and Computer Vision in Smart Aquaculture

Contents

INTRODUCTION

Smart aquaculture is currently one of the sustainable development trends in the aquaculture industry, specifically in the areas of automation and intelligence. Modern intelligent technologies have provided significant advantages in numerous sectors, such as aquaculture, by reducing labor, improving production, and being environmentally friendly. Machine learning is a subset of artificial intelligence (AI) that employs trained algorithm models to identify and acquire characteristics from the data it monitors. As of now, there are numerous studies on the application of machine learning in smart aquaculture, such as the measurement of size, weight, grading, disease detection, and species classification. This chapter offers a comprehensive examination of the advancements in intelligent technology and smart aquaculture. It summarized the methodology, results, and most recent technology that should be employed to advance smart aquaculture over the past nearly decade.

Global aquaculture production has been steadily increasing, surpassing the total live weight by over threefold. The primary species groups, which account for 75% of aquaculture production, are catfish, seaweeds, carps, bivalves, and tilapia. Furthermore, the population of marine species, including fish and crustaceans, has e□panded rapidly during this period. Aquaculture is experiencing a surge in species diversity, with an estimated 40% of the diverse species being fish, shellfish, and algae. These species are cultivated in a variety of water environments, including marine, brackish, and freshwater, on a global scale. Traditional aquaculture is associated with numerous environmental concerns, as well as restricted production and extensive labor requirements. Therefore, smart aquaculture is a concept that aims to promote the sustainable development of the aquaculture industry, increase production, and be environmentally friendly.

In traditional aquaculture, technical procedures consist of water preparation, seed selection, feeding, and care during the rearing process. Aquaculture activities present numerous challenges, including the management of water quality within the aquaculture system. Customarily, individuals collect water samples twice daily, in the morning and afternoon. In certain instances of abrupt changes in water quality in ponds/tanks, it is also impossible to treat the water in a timely manner, and this process requires a significant amount of time. In another instance, individuals are unable to identify diseased fish at an early stage; they must wait until they die or reach the surface before administering treatment. In another instance, the quantity of food that remains in the pond is difficult to estimate with precision, and the quality of the water is impacted by the food. People are required to count each fish before selling them, such as seeds, which requires time and effort. All of the aforementioned challenges have an impact on the profitability of aquaculture. Consequently, the objective of smart aquaculture is to implement smart production methods that address the challenges associated with conventional aquaculture.

In order to implement a smart aquaculture strategy, an environment that is specifically designed to monitor the environmental parameters of the cultured system in real time is integrated with a number of smart devices. The data collected is then used to make decisions automatically. Smart aquaculture is a production mode that is intelligent. It is possible to automate and control it remotely through the application of IoT, big data, artificial intelligence, 5G, cloud computing, and robotics. Conversely, robots can regulate smart aquaculture by overseeing facilities, machinery, and equipment to ensure the successful operation of entire systems.

There are numerous components of smart aquaculture, such as the collection of information through a variety of temperature, dissolved oxygen, humidity, light, and pH sensors to manage the water quality parameters in the aquaculture system; the transmission of the collected data to the control center via communication nodes; the analysis of data and decision-making stored in cloud platforms; the feedback of decision-making to each execution equipment; and the intelligence to operate a system automatically in order to develop aquaculture in a sustainable and efficient manner that is environmentally friendly.

For e□ample, the utilization of AI (artificial intelligence) and IoT (Internet of Things) in aquaculture has been on the rise in order to address the challenges associated with conventional aquaculture. They are employed in a variety of culture systems, including cages, ponds, hatcheries, and breeding, with the objective of achieving a variety of objectives, including the monitoring of water quality, the observation of conditions within cages, ponds, and hatcheries, the optimization of the amount of feed to supply for cultured species and the appropriate time for feeding, the reduction of the frequency of feed supply to cultured systems, and the reduction of labor through the automation of culture systems.

Chrispin et al., (2020) conducted a brief overview of the application of AI in aquaculture, including the use of AI drones, AI feeding devices, AI in shrimp farming, AI smartphone applications, AI in fish processing, AI in open sea fisheries, block chain technology in the shrimp supply chain, and AI in the conservation of endangered fish. Additionally, they reported on the prevention of diseases, fish seed screening, and routine stock checks. Figure 7.1 illustrates the fundamental concept of smart aquaculture. The figure illustrates the deployment of smart devices in aquaculture, as well as the management of the entire aquaculture process, and all data is transmitted via smart phones.

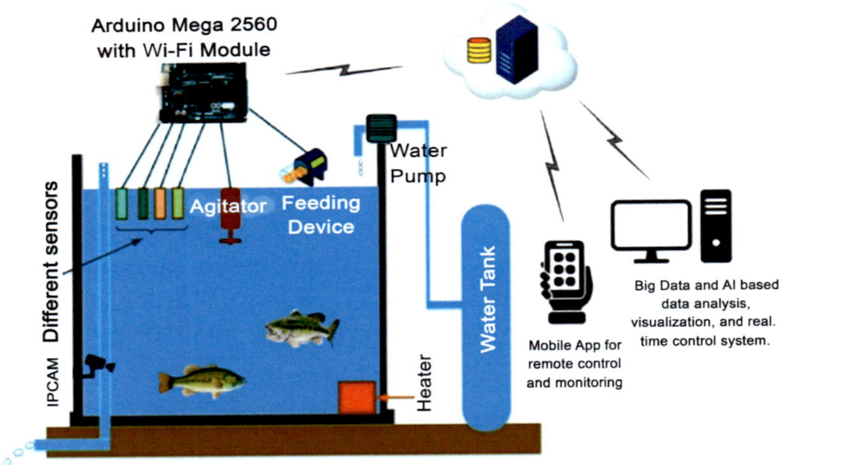

Figure 7.1. *An IoT-based intelligent fish pond.*

Machine learning primarily aims to resolve current problems by employing algorithms and utilizing data to create mathematical models that improve the efficiency of computer systems. Recent aquaculture deployments have incorporated various models, including decision tree (DT), naive Bayes (NB), support vector machine (SVM), artificial neural network (ANN), K-nearest neighbor (KNN), deep learning (DL), and ensemble learning (EL).

Supervised learning is a widely used form of machine learning, alongside unsupervised learning, semi-supervised learning, and reinforcement learning. Using data as a sample after training a machine learning model with the same target values is a widely used practice in supervised learning applications, such as classification and regression. Machine learning has been widely used in aquaculture in recent years. These applications include detecting the biomass of fish, estimating their size and weight, counting them, recognizing different fish species, determining their age, identifying their se□, classifying fish species, analyzing their feeding behavior, studying their group behavior, detecting abnormal behavior, making univariate predictions, and making multivariate predictions. These applications have demonstrated a high level of accuracy.

This chapter intends to assess the present condition of machine learning applications in the field of smart aquaculture. Figure 7.2 provides a comprehensive depiction of smart aquaculture sensors and monitoring systems.

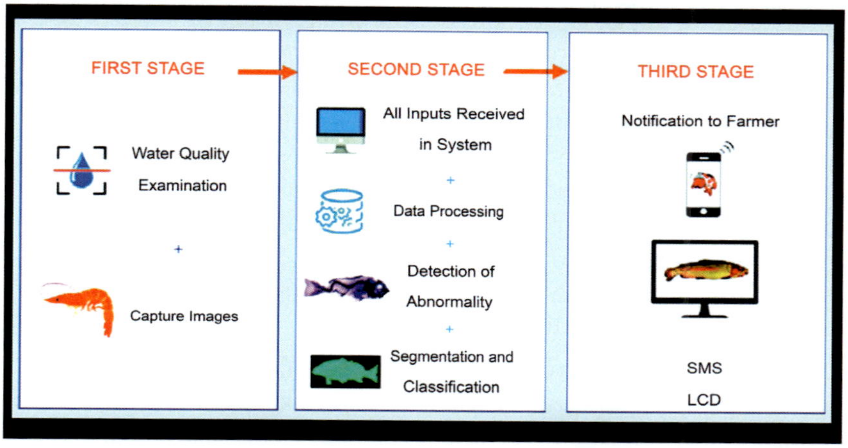

Figure 7.2. *Sensors and monitoring systems used in smart aquaculture.*

7.1. SMART AQUACULTURE

With the advancements in artificial intelligence, smart aquaculture can optimize every stage of the cultivation process for species, including breeding, nursery, and growth. It can also improve other aspects such as managing water resources, ensuring water quality, preparing feed, feeding the species, classifying and grading them, as well as counting and cleaning the cultivation systems. The ultimate objective of developing smart aquaculture is to achieve high aquaculture production that aligns with global demand while also safeguarding the environment.

7.1.1. Monitoring the Quality of Water in Smart Aquaculture

The primary determinant of success and effective management in aquaculture is the quality of water. Various water quality parameters, such as temperature, turbidity, carbon dioxide, pH, alkalinity, ammonia, nitrite, and nitrate, have a significant impact on the survival and growth of cultured species. Temperature, dissolved oxygen, and pH are the most resilient factors. In recent years, the Internet of Things (IoT) has been applied in various fields, including aquaculture. The utilization of IoT in aquaculture has introduced a novel approach to advancing this industry in an environmentally friendly manner. This is achieved through the implementation of intelligent devices that provide real-time monitoring of water conditions, thereby enhancing the working conditions of aquaculture farmers.

The IoT Aquaculture System consists of four primary layers: the Physical

Layer, Monitoring Layer, Virtual Layer, and Connection Protocol, as illustrated in Figure 7.2. The adoption of smart farm technology, also known as intelligent aquaculture, has become an unavoidable trend due to its ability to reduce labor costs, enhance operational efficiency, and ultimately result in higher productivity. Future systems based on the Internet of Things (IoT) could also have the objective of identifying fish diseases and mitigating the impact on production. It is important to acknowledge that while there have been notable advancements in the use of artificial intelligent devices for aquaculture to develop intelligent and highly accurate fish farming, there remain numerous obstacles in fully operating automated systems.

Due to the unique nature of aquaculture activities and its products, labor management in this field carries a high level of risk. This is because there is a requirement for continuous observation, analysis of information, and decision-making throughout the fish farming process. Intelligent equipment, including robotic devices for production, data and information sorting, and energy-saving processing equipment, play a crucial role in monitoring fish farming environments. These technologies will increasingly automate various stages of fish farming operations. These techniques can be utilized in aquaculture to identify fish, estimate their mass, and manage their behavior. Furthermore, it is imperative to develop novel sensors that combine multiple functions within a sensor/multifunctional sensor, ensuring high reliability, a broad range of applications, and a long lifespan. Figure 7.3 depicts an Internet of Things (IoT) aquaculture system.

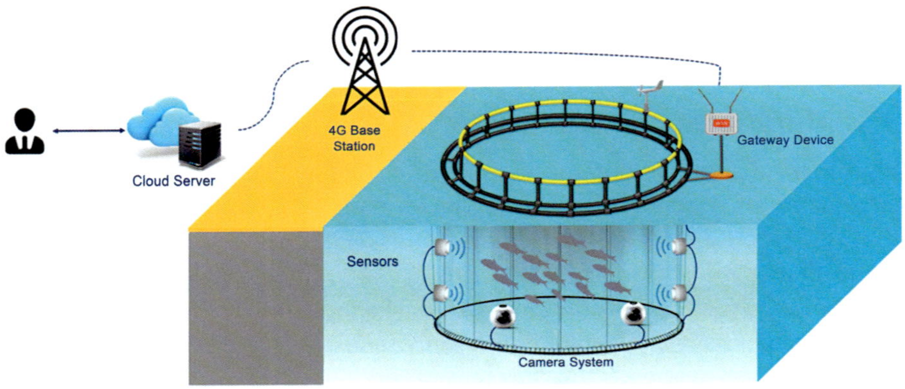

Figure 7.3. *Illustration of aquaculture system that utilizes the internet of things (IoT).*

So far, numerous studies have been conducted on the application of intelligent information technologies, such as artificial intelligence (AI) and the Internet of Things (IoT), in various aspects of aquaculture, including ponds, hatcheries, and cages. Chavan et al., (2018) employed Raspberry Pi to implement real-time monitoring systems in aquaculture. The system included temperature, dissolved oxygen, pH, and ammonia sensors. The gathered data is stored in a cloud computing system and subsequently transmitted to the aqua farmer's mobile device. Kim et al., established a recirculating aquaculture system (RAS) in an intelligent fish farm, equipped with water quality monitoring sensors, MQTT protocol, and a MICOM controller. Al-Hussaini et al., (2018) also concentrate on utilizing fog computing technology and low-cost Raspberry Pi systems to address the current issues in automatic data acquisition and monitoring systems for RAS.

The RAS data collection system, known as RaspDAQ, is created by linking a Raspberry Pi 3 to various components including a temperature sensor (LM35DT), an ADC MCP3002, a water level sensor (HC-SR04), and a RPI camera module. Shin et al., (2020) introduced a sophisticated fish farm design comprising of two tanks: a balancing tank and a recirculating aquaculture system (RAS). The laboratory experiments were conducted at the Future Convergence Technology Research Institute at Busan University of Foreign Studies, while the field e□periments took place at the eel culture farm of Jangsucheon Ltd. on Gangwhado Island.

The water effluence in this system is regulated by a proportional integral derivative (PID) controller, which is linked to a sensor that controls water levels. Additionally, the system includes sensors for water temperature, dissolved oxygen (DO), and potential hydrogen (pH). The Message Queue Telemetry Transport (MQTT) protocol is utilized for remote control and real-time monitoring. The collected large-scale data is stored in laboratory servers. Monirul et al., introduced a water monitoring system that utilizes IoT devices and operates in real-time. The system was constructed utilizing various devices, including sensors for temperature, dissolved oxygen, turbidity, pH, water level, and CO3 gas, as well as an Arduino and an IoT platform.

Nocheski et al., (2018) proposed an enhancement to an existing Internet of Things (IoT) system for monitoring water quality in cultured systems, specifically in ponds. The IoT system comprises sensors capable of measuring various crucial water quality parameters in cultured systems such as tanks, ponds, and cages. These parameters include temperature, light intensity, and water level. Additionally, it includes a compact computer board that gathers data from the sensors and processes it. The analyzed data is then transmitted to the users as a sound or notification. Krishna et al., (2019) implemented an Internet of Things (IoT) system in a fish pond to monitor fish health and water quality. They used an Arduino Uno board, Atmega328 microcontroller, Wi-Fi module, Buzzer, LCD (liquid crystal display), and MIT application. The system collected data from the cloud, which the farmer could access, along with environmental parameters.

Prabhu introduced an Internet of Things (IoT) system designed to monitor and control water quality parameters in a cultured system, such as a lake or pond. This is achieved through the use of sensors, including a temperature sensor, pH sensor, and turbidity sensor. Users can receive and analyze data in the form of SMS on their mobile phone using their preferred language. They can then take various actions to control environmental conditions using an Arduino Nano Board and an ESP8266 wi-fi module. Nguyen et al., (2020) implemented an Internet of Things (IoT) based water quality monitoring system in fish ponds and other aquaculture systems. The main objective was to develop a predictive model for quality indicators.

The author implemented a system where multiple sensors were installed to monitor various water quality parameters, including temperature, salinity, pH, dissolved oxygen (DO), and chemical oxygen demand (COD) indicators, in the fish pond. The software is a cloud-based database offered by Digital Ocean, accessible on both mobile devices and desktop computers. Forecasting techniques employ a model that utilizes an average, stochastic, gradient descent approach. The results obtained from two data sets demonstrate the efficacy of this system, indicating its potential for practical application. Hsu et al., (2020) obtained data on water quality parameters using sensors, including oxidation-reduction potential (ORP), pH, and temperature. In addition, they utilized a real-time monitoring data map to ascertain the current conditions of the pond. The IoT system comprises of Raspberry PI, Arduino UNO, Bluetooth module, multiple sensor modules (ORP, temperature, pH meter), GCM (Google Cloud Messaging), mobile App device's REGID, prediction analysis, and Web Page Management Mode (HTML combined with PHP).

Darmalim et al., (2019) introduced an Internet of Things (IoT) system for the automated monitoring of environmental parameters. The water quality parameters are continuously gathered from the cultured system in real time. The collected data is promptly updated on the web, allowing users to access information through the IoT device. The software is created utilizing a Python framework. By accessing the web application, users can obtain data that accurately represents the environmental conditions. This enables them to promptly identify any occurrences in the pond and take appropriate action accordingly. The electrical components consist of NodeMcuESP8266, ADC Ads1115, a logic converter, a dissolved oxygen and temperature sensor, a pH sensor, a turbidity sensor, a TDS sensor, a DC-DC converter, and a power supply. The system utilizes a Python web development framework for the web application, which is connected to a MySQL database. Darus et al., (2020) suggested the use of simple expectation maximization (EM) clustering to analyze water quality in a catfish pond for monitoring purposes.

Data on various parameters, including total dissolved solids (TDS), water level, turbidity, pH, and temperature, have been gathered using sensors that are linked to a Raspberry Pi 3 Model B. The experiment involves the following tasks: installing the WEKA API on a Raspberry Pi3 Model B (Stretch), installing sensors for total dissolved solids, water level, turbidity, pH, and temperature on the Raspberry Pi3 Model B, and applying a simple expectation maximization clustering technique, as described in the equation. Figure 7.4 depicts the block diagram of an IoT system designed for monitoring water quality, as referenced by sources.

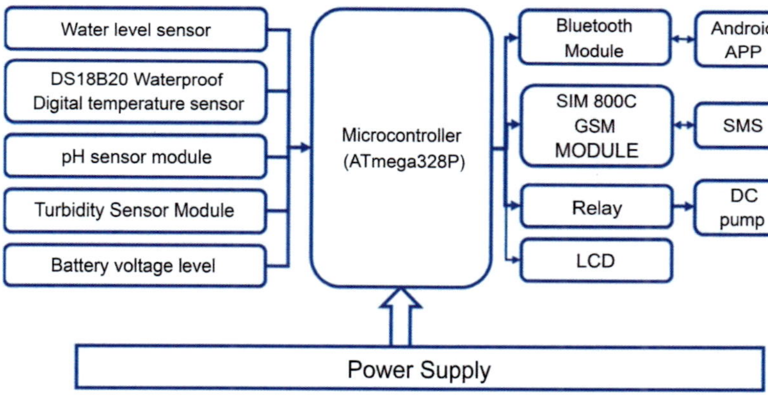

Figure 7.4. *Illustrates the block diagram of an IoT system used for monitoring water quality.*

7.1.2. Feeding Controlling

A primary challenge in aquaculture is the provision of adequate nutrition. As per the conventional approach, farmers will disperse food throughout the pond, or at a specific location within the pond, based on the feeding habits of the cultivated species. Hence, it is imperative to implement IoT in the feeding system to regulate the quantity of feed and enable automated feeding. This approach offers numerous advantages, including manpower conservation and effective management of leftover feed, thereby mitigating water quality pollution in aquaculture. Several studies have been conducted.

Malathi et al., (2018) (developed a highly effective semi-automatic system that promotes the optimal development of aquatic organisms in aquaculture. This study investigates both the feeding system and water quality parameters in a cultured system using sensors like a pH sensor and temperature sensor.

The GSM module is utilized to notify the cultivator whenever the quality parameters exceed the normal range. The feeding system autonomously provides nourishment to the fish according to its specific needs. The system primarily comprises an Arduino Uno, a DS18B20 Temperature Sensor, a pH sensor, and a SIM900A-GSM module. Daud et al., (2020) introduced a water quality system that relies on IoT devices within a smart aquarium. The objective of this study is to ensure that the freshwater in the aquarium tank is maintained at an optimal level to support the habitats and feeding conditions of the fish. This system is specifically designed to utilize MEGA and NodeMCU controllers, with its operation being controlled by a smartphone. Arduinos are utilized in the system's design. The NodeMCU utilizes Wi-Fi communication to establish a connection between the smartphone and the controller, enabling control over the operation. The LCD screen displays all the gathered data regarding the quality of water. Figure 7.5 depicts the comprehensive block diagram of the IoT application's complete system. To summarize, the user has the ability to automate control of the feeding machine using IoT devices.

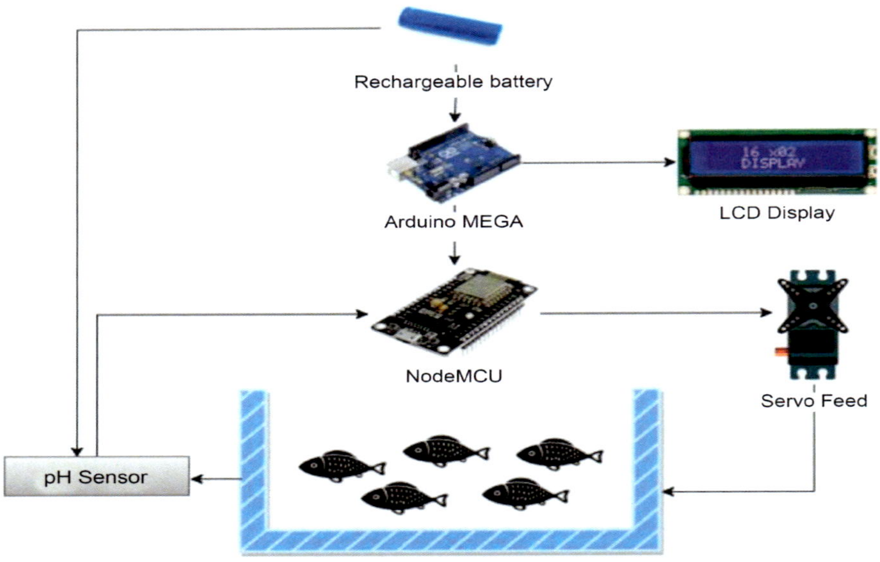

Figure 7.5. *A block diagram of an IoT application's complete system.*

7.2. MACHINE LEARNING

Machine learning is commonly regarded as a subset of artificial intelligence. Algorithmic models are trained to perform specific functions by recognizing and learning samples from data automatically, with accuracy and speed, rather than relying on explicit computer programming expertise. As previously stated, this process can be classified into three categories: supervised, semi-supervised/ unsupervised, and reinforcement learning. The most widely used method is supervised, as shown in Figure 7.6.

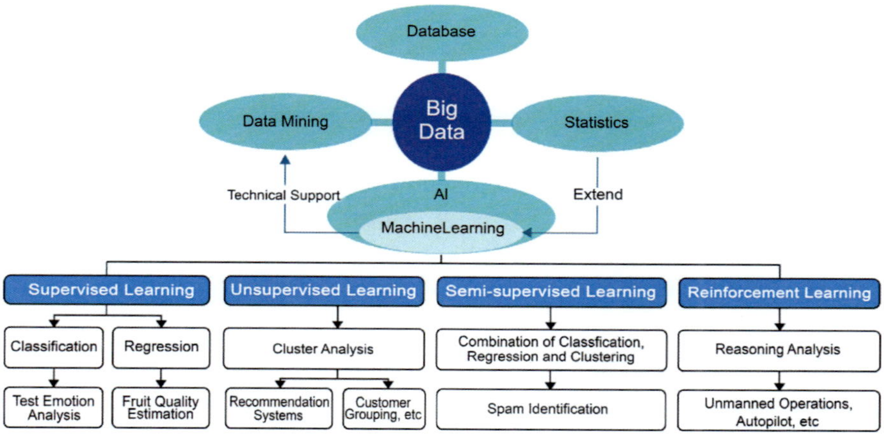

Figure 7.6. *Illustrates the correlation between the four categories of machine learning and their typical applications.*

Supervised learning requires the utilization of an algorithmic system to acquire knowledge by undergoing training with a dataset that has been labeled. Based on these experiences, we can identify and differentiate new data that are dissimilar to the ones used for training. Once the training stages are completed, these algorithms will perform their tasks by testing new data resources/images to determine whether they belong to the target images or not. Unsupervised learning refers to the process of obtaining output data without the use of a training dataset. It bears a strong resemblance to supervised learning. The objective of this form of learning is to categorize the input resources into distinct classifications.

The data lacks labels, but it serves as input for unsupervised learning. The objective of this model is to classify the input data based on its inherent features. The key distinction between supervised learning and unsupervised learning is their robustness. Semi-supervised machine learning refers to a hybrid approach that combines supervised and unsupervised learning. In this approach, the labeled input data is reduced by combining a large number of unlabeled inputs with a small number of labeled inputs.

The four primary tasks in machine learning encompass classification, regression, clustering, and DR (dimensionality reduction). Among these, supervised learning models encompass classification and regression. Clustering is a technique used in unsupervised learning, while dimensionality reduction (DR) can be applied in both supervised and unsupervised learning.

Several machine learning models have been utilized, such as Decision Trees (DT), Support Vector Machines (SVM), Naive Bayes (NB), Artificial Neural Networks (ANN), Deep Learning (DL), Ensemble Learning (EL), and K-Nearest Neighbors (KNN). Every machine learning model possesses unique strengths and weaknesses. According to Zhao et al., (2018), DT (Decision Tree) and SVM (Support Vector Machine) models are generally regarded as effective, but they exhibit a high degree of sensitivity to missing values. Alternatively, NB (Naive Bayes), ANN (Artificial Neural Networks), and KNN (K-Nearest Neighbors) outperform DT (Decision Trees) and SVM (Support Vector Machines) in several aspects, including improved efficiency, reduced sensitivity to missing values, and higher accuracy rates.

7.3. UTILIZATION OF MACHINE LEARNING AND COMPUTER VISION IN AQUACULTURE

Machine vision/computer vision is typically employed as the primary means of quantifying the dimensions of fish. Images captured by a camera connected to a fish measurement system are subsequently transmitted to software for analysis. The resulting data is then compiled and presented as results.

7.3.1. Measure Fish Size

Measuring fish size and grading fish is an essential task in fish farming due to the unique characteristics of fish and the requirement to determine the market size of the fish. Size grading is commonly used during the rearing stage as the most frequent method of grading. Applying machine learning to automate the cultivation of aquatic species in aquaculture can lead to reduced operational and management expenses, while simultaneously improving profitability and the quality of production. Accurate measurement of fish for the market is crucial. This task requires additional time and is complex due to the necessity of removing the abnormal fish. Multiple studies have examined various fish species in the field of aquaculture.

White et al., (2006) introduced a system that utilizes image processing algorithms to accurately determine and identify the dimensions of various fish species. Photographs of various fish species were acquired, including Hippoglossoides platessoides, Solea vulgaris, Microstomus kitt, Pleuronectes platessa, Sebastes marinus, Sebastes mentella, and Platichthys flesus. An Omron PLC, manufactured in Kyoto, Japan, establishes a connection with the main computer and software through an ethernet link, functioning as a regulated mechanical system. Subsequently, the gathered images are transmitted to a computer and subjected to analysis by a software program.

The authors employ hardware and programming methodologies to create fish sorting equipment that can precisely measure fish length and classify fish species under real-time circumstances. The main aim of this study is to accomplish that objective. The result showcases image processing algorithms that are specifically tailored to differentiate between flatfish and roundfish. The length of the fish can be measured with an accuracy of 100%, with a standard deviation of 1.2 mm. Furthermore, the accuracy rate for determining the fish species can reach up to 99.8%. Based on estimates, a single conveyor system can measure up to 30,000 fish per hour.

Costa et al., (2013) conducted an experiment to develop tools that can directly measure the size, identify the sex, and recognize abnormal performance of seabass (Dicentrarchus labrax L.) in online conditions at an aquaculture farm. The seabass samples are obtained from five interbred populations in the European wild. A grand total of 259 fish were meticulously photographed and weighed, with the tiniest fish registering a weight of 0.1 grams. The fish shape is examined through the application of Elliptic Fourier analysis (EFA) on the coordinates of its outline. The original images are transformed through two channels, specifically the G (gray scale) and V (value) channels, within the HSV color space.

The Euclidean distance of each pixel was calculated with respect to the background. The results indicated that this approach is e□ceedingly efficient for online fish categorization and can be widely utilized in different fish farming settings.

Mustafa et al., (2013) introduced a study that utilizes image processing (FLUDI framework) to automatically measure the length of fish, as an alternative to expensive measurement tools. The species Rastrelliger kanagurta and Selar crumenophthalmus were utilized for this experiment. The image displays the ultimate interface of the FLUDI software, which is used to accurately measure the length of fish based on digital images. Two types of cameras were utilized, namely the Pentax camera with a resolution of 8.0 megapixels and the Sony camera with a resolution of 5.0 megapixels.

In this study, the Pentax camera with 8.0 megapixels and the Sony camera with 5.0 megapixels exhibited error rates of 0.74% and 0.19% respectively. The study demonstrated that the FLUDI software possesses significant capability in accurately measuring the true length of fish based on digital images. Nevertheless, the software can be utilized on other objects without requiring any object reference. The fish length obtained from the traditional measuring board was compared to the actual fish length determined from the digital image. The findings indicate that there is a minimal disparity, varying between 0.74% and 2.19%.

Jeong et al., (2018) implemented a vision-based automatic system (VAMS) to measure the total length (TL), body width (BW), height (H), and weight (W) of flatfish without physical contact. This system allows for non-contact measurement of the morphometric characteristics of flatfish. H and W were measured using a laser displacement and a load cell, respectively. Furthermore, a proposed algorithm was utilized to detect and quantify TL (total length) and BW (body width) using a morphological image processing technique. Data can be gathered from various sites, including a research center/institute, a fish market, or a fisheries resource investigation laboratory. Subsequently, all the gathered data is transmitted to the server.

The VAMS consists of a PV-500 vision system from Panasonic in Kadoma, Japan, an Omron ZX-LT030 laser displacement sensor from Kyoto, Japan, a CCD camera with a 25 mm focal length lens and a 2/3-inch CCD size, a 30W LED backlight, a conveyor belt, and a road cell. The findings indicate that the VAMS system is capable of measuring the average total length of flatfish up to 266.844 mm. Additionally, it has the capacity to measure 900 individual fish per hour using a single conveyor. Moreover, it is capable of measuring fish with a total length (TL) of up to 500 mm.

Furthermore, Gerami et al., (2016) employ machine vision to evaluate the mass of salmon fish. The fish specimens were obtained from an aquaculture facility located in Iran. There are a total of 75 living individuals of rainbow trout (Oncorhynchus mykiss). A Canon IXUS 960IS digital camera with a resolution of 12 megapixels (3000 × 4000) was used to capture photographs of the samples from the left side, using the red, green, and blue channels. The camera was positioned directly above the sample at an elevation of 45 cm. The captured fish images can be transferred to a computer and analyzed using MATLAB (Matrix laboratory) version R2009x.

The algorithms achieved a favorable outcome by accurately assessing the weight of fish and producing clear images. Based on this outcome, it can be inferred that machine vision is a highly effective tool for both measuring weight and assessing the visual characteristics of fish. Hence, machine vision can be employed to assess the visual characteristics of fish and appro□imate their weight. It is recommended that this method be employed in fisheries processes in future research. Sanchez-Torres et al., (2018) compiled data from various sources in the literature pertaining to machine vision and subsequently developed a novel framework for this domain, drawing upon the existing information. The utilization of a solitary camera in a controlled environment is an appropriate method for measuring fish, as it allows for better management of sample size and ensures high-quality images. Additionally, a fusion of homomorphic filtering, contrast limited adaptive histogram equalization (CLAHE), and guided filtering was employed for enhancing fish images.

The results demonstrate that polynomial regression is a straightforward and effective technique for measuring the length and weight of fish, particularly for fish in grade 3 and grade 4. Given these findings, this approach is also utilized to determine the quantity of feed provided to fish based on their length and weight.

Sung et al., (2020) conducted a study with the objective of automatically grading flatfish based on size using machine vision, in order to promote effective and stable growth. The grader that was created consists of three main components: a conveyor belt, machine vision, and a sorter.

The fish are transferred onto the conveyor belt for the purpose of quantifying and categorizing. The image processing system identifies fish and quantifies their length. The positioning of the grader is determined according to the length classification.

The machine vision component consists of a camera, LED lights, and a low-cost darkroom. The fish is conveyed from the hopper to the measuring section via the conveyor belt. As a result of the classification, the position of the sorter is controlled. After completing the measurement, the conveyor belt is operated once more. Ultimately, the fish is assessed by the sorter. The findings demonstrated that this system is a precise method for evaluating with e□ceptional efficiency and minimal e□pense.

Typically, machine vision is the primary method used to quantify the dimensions of fish. Images are captured using a camera that is connected to a fish measurement system. These images are then transmitted to a software for analysis, and the results are compiled.

7.3.2. Management of Fish Diseases and Health

Disease is a significant problem that has a profound impact on both the quality and production of aquaculture. In the field of aquaculture, disease arises due to the disruption of various factors originating from the host, pathogen, and environment. The various types of diseases can be classified into two main categories: infectious (parasitic, fungal, bacterial, viral) and non-infectious (environmental, nutritional, and genetic). Due to a lack of expertise, farmers are unable to provide precise treatment for an unexpected fish disease, making it challenging to implement timely and effective treatment measures. The findings indicate that fish disease can rapidly propagate and lead to significant mortality rates, resulting in substantial financial losses for farmers. Machine learning is a method that can be used to promptly identify and notify fish farmers of the presence of disease in their fish.

Epizootic ulcerative syndrome (EUS) is a prevalent disease that has resulted in significant fish mortality rates in several countries, including Australia, India, United Kingdom, Japan, Thailand, and Pakistan. The fungal disease is called Aphanomyces invadans. Malik et al., (2017) conducted a study to develop a method for detecting fish diseases. They combined two techniques, Principal Component Analysis (PCA) and classification using a Neural Network Algorithm (ANN). The input images, specifically fish disease images, were gathered and subsequently subjected to morphology operations, including converting the image to grayscale, eliminating noise, and performing segmentation. The findings indicate that the FAST-PCA-ANN method outperforms the HOG-PCA-ANN technique in terms of both classification accuracy and efficiency (Figure 7.7).

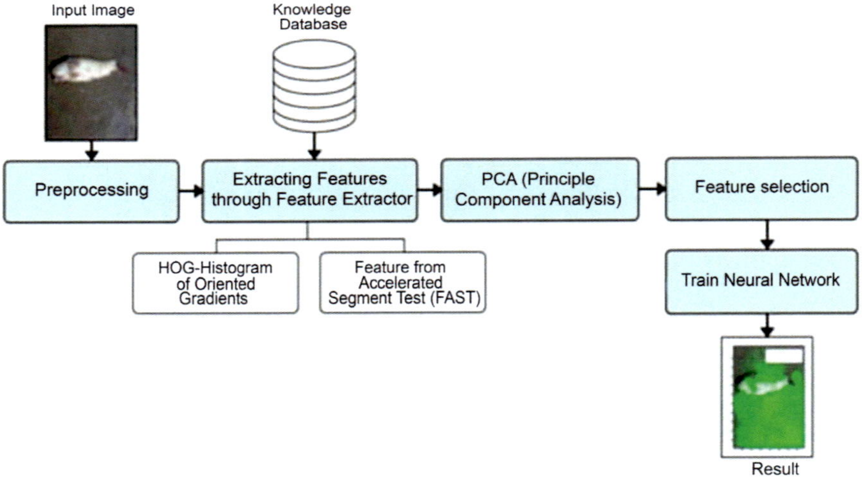

Figure 7.7. *Proposed methodology by Malik et al., (2017).*

Chakravorty et al., (2015) discovered four EUS species that were infected, namely Clarias batrachus, Puntius chola, Labeo bata, and Labeo gonius, in various regions of the Barak Valley, Assam. The researchers employed two machine learning algorithms, namely Principal Component Analysis (PCA) and K-means clustering. The results demonstrated that the algorithm achieved an accuracy rate exceeding 90% for PCA. Ultimately, when it comes to identifying skin color and extracting texture characteristics, HSV proves to be a favorable option. Morphological operations are necessary for detecting fish pathogens and extracting brain tumors (Figures 7.8 and 7.9).

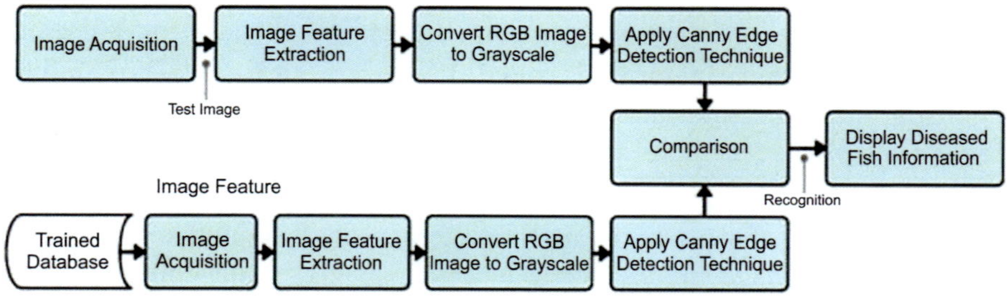

Figure 7.8. *Illustrates the entire process of principal component analysis (PCA).*

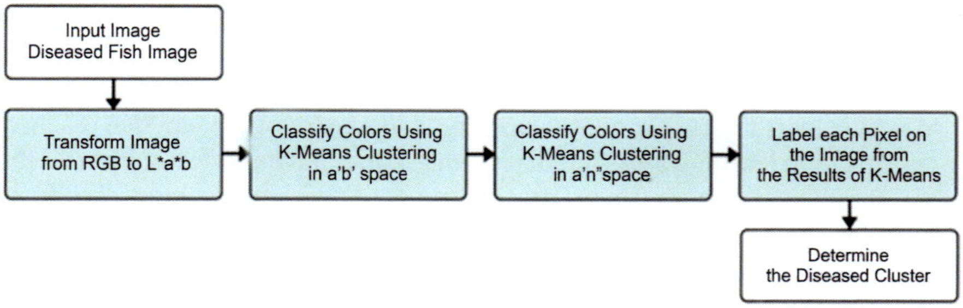

Figure 7.9. *Illustrates the entire process of K-means clustering.*

Divinely et al., (2019) suggest that a probabilistic neural network (PNN) can be used to promptly and efficiently detect fish diseases, specifically EUS ulcers, which is a fungal disease. The procedure is illustrated in Figure 7.10, which outlines the following steps. The input and database images are obtained from different sources and various online resources, respectively. Images undergo preprocessing to mitigate undesired distortions or enhance specific image attributes, which are advantageous for subsequent processing that involves RGB to grayscale conversion. Various extraction techniques have been utilized with CWT (Curvelet Wavelet Transform) to enhance the efficiency of detection. Ultimately, the target fish diseases, including ammonia poisoning, camallanus worm, and dropsy, are categorized. Individuals who have not been infected are identified and isolated.

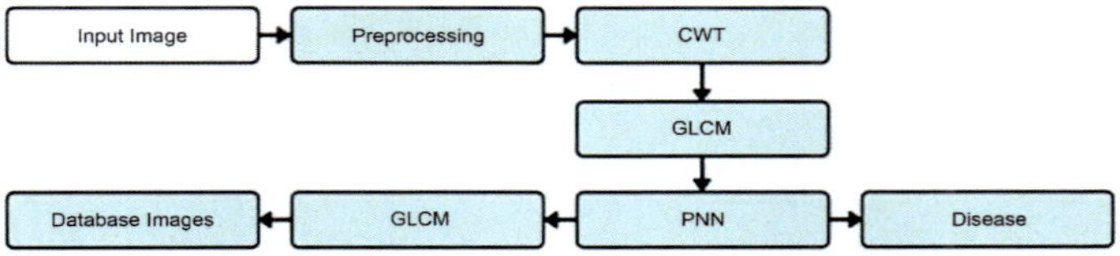

Figure 7.10. *Block diagram of proposed methodology.*

Subsequently, the GLCM (Gray Level Co-Occurrence Matrix) is employed to diminish the dimensionality and retain valuable information. In this study, a supervised learning algorithm called PNN was utilized. The findings demonstrate that the suggested integration of CWT-GLCM-PNN is a highly effective and precise method for detecting fish disease.

Ahmeda et al., (2021) detected diseases in salmon that were raised in a farm. The task is split into two parts: image preprocessing and segmentation. The purpose of image processing is to diminish noise and enhance image clarity. The segmentation process utilizes a Support Vector Machine (SVM) algorithm of machine learning with a kernel function to extract features and classify diseases from images. The findings demonstrated that employing Support Vector Machines (SVM) is a highly effective approach for accurately detecting fish diseases.

Machine learning is commonly employed to accurately detect fish diseases. Various machine learning algorithms include Artificial Neural Networks (ANN), Histogram of Oriented Gradients (HOG), Principal Component Analysis (PCA), and Support Vector Machines (SVM). They can be utilized in conjunction with machine learning algorithms, such as ANN-HOG-PCA, to achieve a high level of accuracy in detecting fish diseases, specifically fungal diseases.

7.3.3. Enumeration

Counting cultured species in aquaculture is essential and can be achieved through the use of computer vision technology, which involves video analysis and image processing. Raman et al., (2016) implemented an image processing technology to develop a counting system specifically designed for quantifying larvae and juveniles in fish hatcheries. The system identifies and analyzes images of larvae and juvenile fish, accurately counting the number of individuals in each image.

The purpose of this is to quantify the population of larvae and juvenile fish using image recognition technology. Larvae and juveniles are captured using a camera. The captured images are transferred to a computer and processed using software, resulting in the analysis outcomes. The results indicate that the installed system achieves an accuracy rate of 82% for larvae and 87% for juveniles.

Cao et al., (2018) employed machine vision to quantify the amount of feed in aquaculture, specifically for fish, in order to effectively monitor and regulate the remaining feed in the system. A current contour refers to a solitary pellet. Using the current contour area, they precisely determined the exact quantity of pellets in a pile. Feed wastage in aquaculture leads to various issues, including reduced profitability, environmental pollution, and negative impacts on the health of the cultured species. Hence, the objective of this study is to quantify the appropriate quantity of feed required for fish cultivation through the utilization of an algorithm. They performed experiments using various levels of water turbidity, feed adhesion rates, and other factors, using more than 100 pellets. Algorithms are an effective solution for accurately counting the number of feeds in the aquaculture system based on real production data. Additionally, it is effective in murky underwater environmental conditions. It is applicable in automatic feeding systems with a high level of precision. This system is capable of being implemented in real-world production scenarios.

Zhang et al., (2020) introduced a model that automates the process of counting fish, replacing the conventional method of artificial sampling. The objective of this system is to accurately quantify the population of fish in cages that are being cultivated in an offshore environment. A hybrid neural network model, consisting of a multi-column convolutional neural network (CNN) and a deep convolutional neural network (DCNN), is employed to accurately and objectively observe, recognize, and quantify the population of fish being cultivated offshore in real-time, without any losses. The findings indicate that the proposed hybrid neural network achieves a counting accuracy rate of 95.06%.

7.3.4. Classification and Identification

Traditional methods relying on visual observation or expensive genetic techniques are reportedly ineffective in achieving accurate results for classification and identification. Hence, machine learning and computer vision are highly effective techniques for rapidly and accurately classifying and identifying fish.

Coz-Rakovac et al., (2009) used machine learning methods to analyze biochemical data and identified three species affected by aquaculture: 120 seabass (Dicentrarchus labrax), 98 sea bream (Sparus aurata L.), and 66 mullets (Mugil spp.). The decision tree yielded the most optimal outcomes among the machine-learning techniques, correctly classifying 210 samples (85.71%) and misclassifying 35 samples (14.29%). Three species are distinguished based on their biochemical characteristics. The study's methodology includes the use of a decision tree and biochemistry analysis. Blood samples were collected to measure AST, TP, TRIG, CHOL, and GLU levels.

Allken et al., (2019) utilized a Deep Vision camera to capture images of the marine stock. These images serve as resources for implementing a deep learning neural network that automates the categorization of different species. The results indicate that a classification accuracy of 94% was attained for blue whiting, Atlantic herring, and Atlantic mackerel. This demonstrates that automatic species classification is a feasible and efficient method. Furthermore, the use of synthetic data proves to be an effective solution for the frequently encountered issue of insufficient training data. The sample fish images are extracted from real ones and placed on a blank background at random positions, sizes, and orientations. The objective of this processing is to generate images that closely resemble photographs taken with deep vision technology. The fish count per image ranges from one to six.

7.4. CONCLUSION

In recent years, the aquaculture industry has witnessed significant growth in the adoption of smart aquaculture techniques, which are characterized by their efficiency, automation, and accuracy. While the implementation of artificial intelligence in aquaculture is advancing rapidly, there remain numerous obstacles to achieving fully automated systems. Given the practices and products involved, the absence of human management poses a significant risk, as there is still a requirement for observation, information analysis, and decision-making throughout the fish farming process.

The utilization of machine learning and computer vision in aquaculture has been actively enhancing the growth of the aquaculture industry in an automated manner, thereby increasing farming efficiency. The utilization of algorithms or images offers the benefit of enhanced precision and swifter performance compared to manual techniques. Nevertheless, the expense associated with machine learning and computer vision is substantial, thus it is advisable to exclusively employ these technologies in extensive agricultural operations that cultivate species with significant economic worth.

This chapter provides an overview of how artificial intelligence, specifically machine learning and computer vision, is being applied to enhance smart aquaculture practices. Scientists and aquaculture practitioners examined the application of machine learning, specifically algorithms or machine vision, in the field of smart aquaculture. The predominant applications involve quantifying the dimensions and mass of cultivated species, diagnosing fish ailments, enumerating fish populations, categorizing and recognizing species, regulating feed intake, and overseeing water purity.

In the future, there is a need for increased accessibility of machine learning and computer vision applications in smart aquaculture. This accessibility should extend beyond hatcheries and land-based farms to include offshore aquaculture systems. Specifically, machine learning and computer vision can be utilized in aquaculture on offshore areas, such as cages, to identify fish diseases, ensure the safety of the cages, and measure the weight and size of the fish. Indeed, it is imperative to detect fish that are afflicted with disease or identify any malfunctions in the cage net system. Typically, farmers frequently employ manual techniques to oversee the entire aquaculture process. For example, diseased fish can be identified by their lethargic swimming near the water's surface or by requiring individuals to enter the enclosure for inspection. Thus, the utilization of machine learning/vision in conjunction with an underwater camera holds significant worth. This system possesses the ability to identify fish ailments and effectively oversee the cage, weight, and size of fish in a direct and uninterrupted manner. It is envisioned as a prospective model system for offshore cage culture.

REFERENCES

1. Adegboye, M. A., Aibinu, A. M., Kolo, J. G., Aliyu, I., Folorunso, T. A., & Lee, S. (2020). Incorporating intelligence in fish feeding system for dispensing feed based on fish feeding intensity. *IEEE Access, 8*, 91948–91960. https://doi.org/10.1109/ACCESS.2020.2991600.
2. Ahmeda, S., Aurpaa, T. T., & Azada, A. K. (2021). Fish disease detection using image-based machine learning technique in aquaculture. *Journal of King Saud University–Computer and Information Sciences*, 1–15. https://doi.org/10.1016/j.jksuci.2021.04.003.
3. Al-Hussaini, K., Zainol, S. M., Ahmed, R. B., & Daud, S. (2018). IoT monitoring and automation data acquisition for recirculating aquaculture system using fog computing. *Journal of Computer Hardware Engineering, 1*. https://doi.org/10.1155/2018/9137154.
4. Allken, V., Handegard, N. O., Rosen, S., Schreyeck, T., Mahiout, T., & Malde, K. (2019). Fish species identification using a convolutional neural network trained on synthetic data. *ICES Journal of Marine Science, 76*, 342–349. https://doi.org/10.1093/icesjms/fsx244.
5. Allken, V., Handegard, N. O., Rosen, S., Schreyeck, T., Mahiout, T., & Malde, K. (2019). Fish species identification using a convolutional neural network trained on synthetic data. *ICES Journal of Marine Science, 76*, 342–349. https://doi.org/10.1093/icesjms/fsz006.
6. Barulin, N. V. (2017). E□ternal se□ specific signs in the structure of derivatives of sterlet (*Acipenser ruthenus*, Linnaeus, 1758) corium. *Acta Biologica Universitatis Daugavpiliensis, 17*, 9–17. https://doi.org/10.15181/abuds.v17i0.1711.
7. Barulin, N. V. (2019). Using machine learning algorithms to analyse the scute structure and se□ identification of sterlet *Acipenser ruthenus* (Acipenseridae). *Aquaculture Research, 50*, 2810–2825. https://doi.org/10.1111/are.14105.
8. Barzegar, R., Aalami, M. T., & Adamowski, J. (2020). Short-term water quality variable prediction using a hybrid CNN—LSTM deep learning model. *Stochastic Environmental Research and Risk Assessment, 34*, 415–433. https://doi.org/10.1007/s00477-019-01715-0.
9. Cai, K., Miao, X., Wang, W., Pang, H., Liu, Y., & Song, J. (2020). A modified YOLOv3 model for fish detection based on MobileNetv1 as backbone. *Aquacultural Engineering, 91*, 102117. https://doi.org/10.1016/j.aquaeng.2020.102117.
10. Cao, J., & Xu, L. (2018). Research on counting algorithm of residual feeds

in aquaculture based on machine vision. In *Proceedings of the 2018 IEEE 3rd International Conference on Image, Vision and Computing (ICIVC)* (pp. 498–503). IEEE. https://doi.org/10.1109/ICIVC.2018.00087.

11. Cao, X., Liu, Y., Wang, J., Liu, C., & Duan, Q. (2020). Prediction of dissolved oxygen in pond culture water based on K-means clustering and gated recurrent unit neural network. *Aquacultural Engineering, 91*, 102122. https://doi.org/10.1016/j.aquaeng.2020.102122.
12. Chakravorty, H., Paul, R., & Das, P. (2015). Image processing technique to detect fish disease. *International Journal of Computer Science and Security (IJCSS), 9*, 121–131. https://doi.org/10.2316/Journal.206.2015.9.206-240.
13. Chartrand, G., Cheng, P. M., Vorontsov, E., Drozdzal, M., Turcotte, S., Pal, C. J., Kadoury, S., & Tang, A. (2017). Deep learning: A primer for radiologists. *Radiographics, 37*, 2113–2131. https://doi.org/10.1148/rg.2017170077.
14. Chavan, M. S., Patil, V. P., Chavan, S., Sana, S., & Shinde, C. (2018). Design and implementation of IoT-based real-time monitoring system for aquaculture using Raspberry Pi. *International Journal of Recent Innovations in Computing and Communication, 6*, 159–161. https://doi.org/10.21742/ijricc.2018.6.28.
15. Chhabra, H. S., Srivastava, A. K., & Nijhawan, R. (2020). A hybrid deep learning approach for automatic fish classification. In *Proceedings of ICETIT 2019* (pp. 185–194). Springer. https://doi.org/10.1007/978-3-030-35678-2_18.
16. Costa, C., Antonucci, F., Boglione, C., Menesatti, P., Vandeputte, M., & Chatain, B. (2013). Automated sorting for size, sex and skeletal anomalies of cultured seabass using external shape analysis. *Aquacultural Engineering, 52*, 58–64. https://doi.org/10.1016/j.aquaeng.2012.09.001.
17. Coz-Rakovac, R., Topic Popovic, N., Smuc, T., Strunjak-Perovic, I., & Jadan, M. (2009). Classification accuracy of algorithms for blood chemistry data for three aquaculture-affected marine fish species. *Fish Physiology and Biochemistry, 35*, 641–647. https://doi.org/10.1007/s10695-008-9271-2.
18. Coz-Rakovac, R., Topic Popovic, N., Smuc, T., Strunjak-Perovic, I., & Jadan, M. (2009). Classification accuracy of algorithms for blood chemistry data for three aquaculture-affected marine fish species. *Fish Physiology and Biochemistry, 35*, 641–647. https://doi.org/10.1007/s10695-008-9276-2.
19. Darmalim, U., Darmalim, F., Darmalim, S., Ahmad, A., Budiarto, A., Mahesworo, B., & Pardamean, B. (2019). IoT solution for intelligent pond

monitoring. In *Proceedings of the 3rd International Conference on Eco Engineering Development* (Vol. 426, pp. 1–10). IOP Publishing. https://doi.org/10.1088/1755-1315/426/1/012014.

20. Darus, F., Ayob, M. Z., Malaysia, G., Hassan, H. A., Ridzuan, I. S., Razak, M. R. A., Baharuddin, H., & Rahman, A. H. A. (2020). Potential of using Internet of Things (IoT) for water quality monitoring in aquaculture: A case study in freshwater catfish culture in Rawang, Selangor, Malaysia. *Test Engineering and Management, 83*, 2163–2169.
21. Daud, A. K. P. M., Sulaiman, N. A., Yusof, Y. W. M., & Kassim, M. (2020). An IoT-based smart aquarium monitoring system. In *Proceedings of the 2020 IEEE 10th Symposium on Computer Applications & Industrial Electronics (ISCAIE)* (pp. 277–282). IEEE. https://doi.org/10.1109/ISCAIE49889.2020.9129082.
22. Deep, B., & Dash, R. (2019). Underwater fish species recognition using deep learning techniques. In *Proceedings of the 2019 6th International Conference on Signal Processing and Integrated Networks (SPIN)* (pp. 665–669). IEEE. https://doi.org/10.1109/SPIN.2019.8711855.
23. Divinely, S. J., Sivakami, K., & Jayaraj, V. (2019). Fish diseases identification and classification using machine learning. *International Journal of Advanced Research in Basic Engineering Sciences and Technology (IJARBEST), 5*, 46–51.
24. Dzulqornain, M. I., Rasyid, M. U. H. A., & Sukaridhoto, S. (2017). Design and development of smart aquaculture system based on IFTTT model and cloud integration. In *Proceedings of the 3rd International Conference on Electrical Systems, Technology and Information (ICESTI 2017)* (Vol. 164). https://doi.org/10.1109/ICESTI.2017.8091170.
25. Erickson, B. J., Korfiatis, P., Akkus, Z., & Kline, T. L. (2017). Machine learning for medical imaging. *Radiographics, 37*, 505–515. https://doi.org/10.1148/rg.2017160168.
26. Fernandes, A. F. A., Turra, E. M., de Alvarenga, E. R., Passafaro, T. L., Lopes, F. B., Alves, G. F. O., Singh, V., & Rosa, G. J. M. (2020). Deep learning image segmentation for e□traction of fish body measurements and prediction of body weight and carcass traits in Nile tilapia. *Computers and Electronics in Agriculture, 170*, 105274. https://doi.org/10.1016/j.compag.2020.105274.
27. Fijani, E., Barzegar, R., Deo, R., Tziritis, E., & Skordas, K. (2019). Design and implementation of a hybrid model based on two-layer decomposition method coupled with extreme learning machines to support real-time environmental monitoring of water quality parameters.

Science of the Total Environment, 648, 839–853. https://doi.org/10.1016/j.scitotenv.2018.08.257.

28. França Albuquerque, P. L., Garcia, V., Da Silva Oliveira, A., Lewandowski, T., Detweiler, C., Gonçalves, A. B., Costa, C. S., Naka, M. H., & Pistori, H. (2019). Automatic live fingerlings counting using computer vision. *Computers and Electronics in Agriculture, 167*, 105015. https://doi.org/10.1016/j.compag.2019.105015.
29. Garcia, R., Prados, R., Quintana, J., Tempelaar, A., Gracias, N., Rosen, S., Vågstøl, H., & Løvall, K. (2019). Automatic segmentation of fish using deep learning with application to fish size measurement. *ICES Journal of Marine Science, 77*, 1354–1366. https://doi.org/10.1093/icesjms/fsz054.
30. Gerami, M. H., Safiyari, H., Jafari, A., & Mousavi-Nasab, M. (2016). Application of machine vision to assess weight of fish (Case study: *Oncorhynchus mykiss*). *Iranian Journal of Fisheries Sciences, 15*, 575–584. https://doi.org/10.22092/ijfs.2016.118471.
31. Hamid, M. S., Wahab, M. A. A., Abdullah, R., Gani, S. F. B. A., & Hamzah, R. A. (2019). Development of water quality for smart aquaculture system. *Journal of Engineering and Applied Sciences, 14*, 2840–2847. https://doi.org/10.36478/jeasci.2019.2840.2847.
32. Han, F., Zhu, J., Liu, B., Zhang, B., & Xie, F. (2020). Fish shoals' behavior detection based on convolutional neural network and spatiotemporal information. *IEEE Access, 8*, 126907–126926. https://doi.org/10.1109/ACCESS.2020.3007716.
33. Harish, B., Manigandan, K., Harishankar, N., & Ram, T. (2018). Smart aquaculture fish feeding and water quality monitoring. In *Proceedings of the Conference on Electronics, Information and Communication Systems (CELICS'18)*.
34. Hsu, W. C., Chao, P. Y., Wang, C. S., Hsieh, J. C., & Huang, W. (2020). Application of regression analysis to achieve a smart monitoring system for aquaculture. *Information, 11*, 387. https://doi.org/10.3390/info11100387.
35. Hu, W., Wu, H., Zhang, Y., Zhang, S., & Lo, C. (2020). Shrimp recognition using ShrimpNet based on convolutional neural network. *Journal of Ambient Intelligence and Humanized Computing*. https://doi.org/10.1007/s12652-020-02758-2.
36. Imai, T., Arai, K., & Kobayashi, T. (2019). Smart aquaculture system: A remote feeding system with smartphones. In *Proceedings of the 2019 IEEE 23rd International Symposium on Consumer Technologies (ISCT)* (pp. 93–96). IEEE. https://doi.org/10.1109/ISCT.2019.00028.
37. Jalal, A., Salman, A., Mian, A., Shortis, M., & Shafait, F. (2020). Fish

detection and species classification in underwater environments using deep learning with temporal information. *Ecological Informatics, 57*, 101088. https://doi.org/10.1016/j.ecoinf.2020.101088.

38. Jeong, S. J., Yang, Y. S., Lee, K. H., Kang, J. G., & Lee, D. G. (2018). Vision-based automatic system for non-contact measurement of morphometric characteristics of flatfish. *Journal of Electrical Engineering & Technology, 8*, 1194–1201. https://doi.org/10.5370/JEET.2018.8.4.1194.
39. Jia, B., & Zhang, M. (2020). Multi-dimensional classification via kNN feature augmentation. *Pattern Recognition, 106*, 107423. https://doi.org/10.1016/j.patcog.2020.107423.
40. Jordan, M. I., & Mitchell, T. M. (2015). Machine learning: Trends, perspectives, and prospects. *Science, 349*, 255–260. https://doi.org/10.1126/science.aaa8415.
41. Kassem, T., Shahrour, I., El Khattabi, J., & Raslan, A. (2021). Smart and sustainable aquaculture farms. *Sustainability, 13*, 685. https://doi.org/10.3390/su130100685.
42. Keshtegar, B., Keshtegar, B., Heddam, S., & Heddam, S. (2018). Modeling daily dissolved o□ygen concentration using modified response surface method and artificial neural network: A comparative study. *Neural Computing and Applications, 30*, 2995–3006. https://doi.org/10.1007/s00500-017-2766-8.
43. Kim, B. J., & Shin, K. J. (2018). Development of smart fish farming system based on IoT using wasted warm water energy. *Journal of the Institute of Control, Robotics and Systems, 24*, 155–163. https://doi.org/10.53026/jicrs.2018.24.2.155.
44. Kim, S., Alizamir, M., Zounemat-Kermani, M., Kisi, O., & Singh, V. P. (2020). Assessing the biochemical oxygen demand using neural networks and ensemble tree approaches in South Korea. *Journal of Environmental Management, 270*, 110834. https://doi.org/10.1016/j.jenvman.2020.110834.
45. Kotsiantis, S. B. (2007). Supervised machine learning: A review of classification techniques. *Informatica, 31*, 249–268.
46. Krishna, P. C., Rao, M. V., & Reddy, V. R. (2019). An IoT based fish health monitoring system in aquaculture farming. *International Journal of Innovative Technology and Exploring Engineering (IJITEE), 8*, 1578–1584. https://doi.org/10.35940/ijitee.A6595.078919.
47. Le, J., & Xu, L. (2017). An automated fish counting algorithm in aquaculture based on image processing. In *Proceedings of the 2016 International Forum on Mechanical, Control and Automation (IFMCA 2016)* (pp. 358–

366). Atlantis Press. https://doi.org/10.2991/ifmca-16.2016.63.

48. LeCun, Y., Bengio, Y., & Hinton, G. (2015). Deep learning. *Nature, 521*, 436–444. https://doi.org/10.1038/nature14539.
49. Li, D., & Li, C. (2020). Intelligent aquaculture. *Journal of the World Aquaculture Society, 51*, 808–814. https://doi.org/10.1111/jwas.12704.
50. Li, D., Miao, Z., Peng, F., Wang, L., Hao, Y., Wang, Z., Chen, T., Li, H., & Zheng, Y. (2021). Automatic counting methods in aquaculture: A review. *Journal of the World Aquaculture Society, 52*, 269–283. https://doi.org/10.1111/jwas.12735.
51. Li, P., Hua, P., Gui, D., Niu, J., Pei, P., Zhang, J., & Krebs, P. (2020). A comparative analysis of artificial neural networks and wavelet hybrid approaches to long-term toxic heavy metal prediction. *Scientific Reports, 10*, 13439. https://doi.org/10.1038/s41598-020-70387-2.
52. Liu, L., Lu, H., Cao, Z., & Xiao, Y. (2018). Counting fish in sonar images. In *Proceedings of the 2018 25th IEEE International Conference on Image Processing (ICIP)* (pp. 3189–3193). IEEE. https://doi.org/10.1109/ICIP.2018.8451248.
53. Lloyd Chrispin, C., Jothiswaran, V. V., Velumani, T., Agnes Daney Angela, S., & Jayaraman, R. (2020). Application of artificial intelligence in fisheries and aquaculture. *Biotica Research Today, 2*, 499–502. https://doi.org/10.37862/bri.2020.0204.
54. Lu, H., & Ma, X. (2020). Hybrid decision tree-based machine learning models for short-term water quality prediction. *Chemosphere, 249*, 126–169. https://doi.org/10.1016/j.chemosphere.2020.126169.
55. Ma, Y. (2012). *Ensemble Machine Learning: Methods and Applications*. Springer Science & Business Media.
56. Metian, M., Troell, M., Christensen, V., Steenbeek, J., & Pouil, S. (2020). Mapping diversity of species in global aquaculture. *Reviews in Aquaculture, 12*, 1090–1100. https://doi.org/10.1111/raq.12364.
57. Moen, E., Handegard, N. O., Allken, V., Albert, O. T., Harbitz, A., & Malde, K. (2018). Automatic interpretation of otoliths using deep learning. *PLoS ONE, 13*, e204713. https://doi.org/10.1371/journal.pone.0204713.
58. Monirul, I., Mohammod, A. K., & Farzana, I. J. (2019). Aqua fishing monitoring system using IoT devices. *International Journal of Innovative Science, Engineering & Technology (IJISET), 6*, 108–114.
59. Monkman, G. G., Hyder, K., Kaiser, M. J., & Vidal, F. P. (2019). Using machine vision to estimate fish length from images using regional convolutional neural networks. *Methods in Ecology and Evolution, 10*,

2045–2056. https://doi.org/10.1111/2041-210X.13210.

60. Moore, M. M., Slonimsky, E., Long, A. D., Sze, R. W., & Iyer, R. S. (2019). Machine learning concepts, concerns, and opportunities for a pediatric radiologist. *Pediatric Radiology, 49*, 509–516. https://doi.org/10.1007/s00247-018-4294-3.

61. Morimoto, T., Zin, T. T., & Itami, T. (2018). A study on abnormal behavior detection of infected shrimp. In *Proceedings of the 2018 IEEE 7th Global Conference on Consumer Electronics (GCCE)* (pp. 291–292). IEEE. https://doi.org/10.1109/GCCE.2018.8574485.

62. Mustafa, M., Zaidi, M. Z., Shafry, M. M. R., Ismail, M. A., & Norhaida, A. (2013). FLUDI: Using digital images for measuring fish length. *Galaxea: Journal of Coral Reef Studies, 15*, 101–106. https://doi.org/10.3755/galaxea.15.101.

63. Naylor, R. L., Hardy, R. W., & Buschmann, A. H. (2021). A 20-year retrospective review of global aquaculture. *Nature, 591*, 551–563. https://doi.org/10.1038/s41586-020-0310-9.

64. Nguyen, T. N., Tran, T. H., & Nguyen, C. N. (2020). A forecasting model for monitoring water quality in aquaculture and fisheries IoT systems. In *Proceedings of the 2020 International Conference on Advanced Computing and Applications (ACOMP)* (pp. 165–169). Quy Nhon, Vietnam. https://doi.org/10.1109/ACOMP51583.2020.00036.

65. Nocheski, S., & Naumoski, A. (2018). Water monitoring IoT system for fish farming ponds. *Industry 4.0, 3*, 77–79.

66. Ordonez, A., Eikvil, L., Salberg, A., Harbitz, A., Murray, S. M., & Kampffmeyer, M. C. (2020). Explaining decisions of deep neural networks used for fish age prediction. *PLoS ONE, 15*, e235013. https://doi.org/10.1371/journal.pone.0235013.

67. Petrellis, N. (2021). Measurement of fish morphological features through image processing and deep learning techniques. *Applied Sciences, 11*, 4416. https://doi.org/10.3390/app11104416.

68. Prabhu, T. (2019). Automated enhancement of aquaculture species growth by observing the water quality using IoT. *International Research Journal of Multidisciplinary Technovation (IRJMT), 1*, 156–164.

69. Pramunendar, R. A., Wibirama, S., & Santosa, P. I. (2019). Fish classification based on underwater image interpolation and back-propagation neural network. In *Proceedings of the 2019 5th International Conference on Science and Technology (ICST)* (Vol. 1, pp. 1–6). Yogyakarta, Indonesia. https://doi.org/10.1109/ICST.2019.8899323.

70. Rahman, A., Akter, S., Khan, M. M., & Rhaman, K. (2019). Relation between aquaculture with fish disease and health management: A review note. *Bangladesh Journal of Fisheries, 31*, 253–260. https://doi.org/10.3329/bjf.v31i1.41358.
71. Rahman, M., Bapery, C., Hossain, M. J., Hassan, Z., Hossain, G. M. J., & Islam, M. (2020). Internet of Things (IoT) based water quality monitoring system. *International Journal of Multidisciplinary Current Educational Research (IJMCER), 2*(4), 168–180.
72. Raman, V., Perumal, S., Navaratnam, S., & Fazilah, S. (2016). Computer-assisted counter system for larvae and juvenile fish in Malaysian fishing hatcheries by machine learning approach. *Journal of Computer Science, 11*, 423–431. https://doi.org/10.3844/jcssp.2016.423.431.
73. Rashid, M., Nayan, A. A., Simi, S. A., Saha, J., Rahman, O., & Kibria, M. G. (2021). IoT based smart water quality prediction for biofloc aquaculture. *International Journal of Advanced Computer Science and Applications (IJACSA), 12*(6), 56–62. https://doi.org/10.14569/IJACSA.2021.0120608.
74. Rauf, H. T., Lali, M. I. U., Zahoor, S., Shah, S. Z. H., Rehman, A. U., & Bukhari, S. A. C. (2019). Visual features based automated identification of fish species using deep convolutional neural networks. *Computers and Electronics in Agriculture, 167*, 105075. https://doi.org/10.1016/j.compag.2019.105075.
75. Ren, Q., Wang, X., Li, W., Wei, Y., & An, D. (2020). Research of dissolved oxygen prediction in recirculating aquaculture systems based on deep belief network. *Aquacultural Engineering, 90*, 102085. https://doi.org/10.1016/j.aquaeng.2020.102085.
76. Rum, S. N. M., & Nawawi, F. A. Z. (2021). FishDeTec: A fish identification application using image recognition approach. *International Journal of Advanced Computer Science and Applications (IJACSA), 12*(3), 102–106. https://doi.org/10.14569/IJACSA.2021.0120314.
77. Sanchez-Torres, G., Ceballos-Arroyo, A., & Robles-Serrano, S. (2018). Automatic measurement of fish weight and size by processing underwater hatchery images. *Engineering Letters, 26*, 461–472.
78. Sharma, D., & Kumar, R. (2021). Smart aquaculture: Integration of sensors, biosensors, and artificial intelligence. In *Biosensors in Agriculture: Recent Trends and Future Perspectives. Concepts and Strategies in Plant Sciences* (pp. 1–25). Springer. https://doi.org/10.1007/978-3-030-66131-6_1.
79. Shin, K. J. (2020). Development of a mobile integrated control app for smart fish farms based on the IoT. *IEIE Transactions on Smart Processing and Computing, 9*, 142–150. https://doi.org/10.5573/IEIESPC.2020.9.2.142.

80. Shubhaker, B., Prasad, D., Manikanta, S., & Ba, P. A. (2020). Automated fish farm aquaculture monitoring system using IoT. *Journal of Engineering Science, 11*, 1170–1172.

81. Siddiqui, S. A., Salman, A., Malik, M. I., Shafait, F., Mian, A., Shortis, M. R., & Harvey, E. S. (2018). Automatic fish species classification in underwater videos: Exploiting pre-trained deep neural network models to compensate for limited labelled data. *ICES Journal of Marine Science, 75*(1), 374–389. https://doi.org/10.1093/icesjms/fsx199.

82. Sivabalan, K. N., Anandkumar, V., & Balakrishnan, S. (2020). IoT based smart farming for effective utilization of water and energy. *International Journal of Advanced Science and Technology, 29*, 2496–2500.

83. Sung, H. J., Park, M. K., & Choi, J. W. (2020). Automatic grader for flatfishes using machine vision. *International Journal of Control, Automation, and Systems, 18*, 3073–3082. https://doi.org/10.1007/s12555-019-0238-7.

84. Ta, X., & Wei, Y. (2018). Research on a dissolved oxygen prediction method for recirculating aquaculture systems based on a convolutional neural network. *Computers and Electronics in Agriculture, 145*, 302–310. https://doi.org/10.1016/j.compag.2018.01.013.

85. Tang, Q., Qiu, W., & Zhou, Y. (2020). Classification of comple□ power quality disturbances using optimized S-transform and kernel SVM. *IEEE Transactions on Industrial Electronics, 67*, 9715–9723. https://doi.org/10.1109/TIE.2020.2999744.

86. Tharwat, A., Hemedan, A. A., Hassanien, A. E., & Gabel, T. (2018). A biometric-based model for fish species classification. *Fisheries Research, 204*, 324–336. https://doi.org/10.1016/j.fishres.2018.03.007.

87. Villon, S., Mouillot, D., Chaumont, M., Darling, E. S., Subsol, G., Claverie, T., & Villéger, S. (2018). A deep learning method for accurate and fast identification of coral reef fishes in underwater images. *Ecological Informatics, 48*, 238–244. https://doi.org/10.1016/j.ecoinf.2018.06.009.

88. Wang, C., Li, Z., Wang, T., Xu, X., Zhang, X., & Li, D. (2021). Intelligent fish farm - the future of aquaculture. *Aquaculture International, 29*, 2681–2711. https://doi.org/10.1007/s10499-021-00616-w.

89. Webb, M. A. H., Van Eenennaam, J. P., Crossman, J. A., & Chapman, F. A. (2019). A practical guide for assigning sex and stage of maturity in sturgeons and paddlefish. *Journal of Applied Ichthyology, 35*, 169–186. https://doi.org/10.1111/jai.13922.

90. White, D. J., Svellingen, C., & Strachan, N. J. C. (2006). Automated measurement of species and length of fish by computer vision. *Fisheries Research, 80*, 203–210. https://doi.org/10.1016/j.fishres.2006.03.011.

91. Xu, F., Pan, Z., & Xia, R. (2020). E-commerce product review sentiment classification based on a naïve Bayes continuous learning framework. *Information Processing & Management, 57*, 102221. https://doi.org/10.1016/j.ipm.2020.102221.
92. Xu, W., & Matzner, S. (2018). Underwater fish detection using deep learning for water power applications. In *Proceedings of the 2018 International Conference on Computational Science and Computational Intelligence (CSCI)* (pp. 313–318). Las Vegas, NV, USA. https://doi.org/10.1109/CSCI.2018.00065.
93. Yang, L., Liu, Y., Yu, H., Fang, X., Song, L., Li, D., & Chen, Y. (2020). Computer vision models in intelligent aquaculture with emphasis on fish detection and behavior analysis: A review. *Archives of Computational Methods in Engineering, 28*, 2785–2816. https://doi.org/10.1007/s11831-020-09459-7.
94. Yang, X., Zhang, S., Liu, J., Gao, Q., Dong, S., & Zhou, C. (2020). Deep learning for smart fish farming: Applications, opportunities, and challenges. *Reviews in Aquaculture*. https://doi.org/10.1111/raq.12503.
95. Zhakov, A., Zhu, H., Siegel, A., Rank, S., Schmidt, T., Fienhold, L., & Hummel, S. (2020). Application of ANN for fault detection in overhead transport systems for semiconductor fab. *IEEE Transactions on Semiconductor Manufacturing, 33*, 337–345. https://doi.org/10.1109/TSM.2020.3015501.
96. Zhang, L., Wang, J., & Duan, Q. (2020). Estimation for fish mass using image analysis and neural network. *Computers and Electronics in Agriculture, 173*. https://doi.org/10.1016/j.compag.2020.105380.
97. Zhang, S., Yang, X., Wang, Y., Zhao, Z., Liu, J., Liu, Y., Sun, C., & Zhou, C. (2020). Automatic fish population counting by machine vision and a hybrid deep neural network model. *Animals, 10*, 364. https://doi.org/10.3390/ani10020364.
98. Zhang, Y., Fitch, P., & Thorburn, P. J. (2020). Predicting the trend of dissolved oxygen based on the kPCA-RNN model. *Water, 12*, 585. https://doi.org/10.3390/w12020585.
99. Zhao, J., Bao, W., Zhang, F., Zhu, S., Liu, Y., Lu, H., Shen, M., & Ye, Z. (2018). Modified motion influence map and recurrent neural network-based monitoring of the local unusual behaviors for fish school in intensive aquaculture. *Aquaculture, 493*, 165–175. https://doi.org/10.1016/j.aquaculture.2018.04.037.
100. Zhao, S., Zhang, S., Liu, J., Wang, H., Zhu, J., Li, D., & Zhao, R. (2021). Application of machine learning in intelligent fish aquaculture:

A review. *Aquaculture, 540*, 1–19. https://doi.org/10.1016/j.aquaculture.2021.736751.

101. Zhou, C., Sun, C., Lin, K., Xu, D., Guo, Q., Chen, L., & Yang, X. (2018). Handling water reflections for computer vision in aquaculture. *Transactions of the ASABE, 61*, 469–479. https://doi.org/10.13031/trans.12693.

Chapter 8

Current Practices in Organic Aquaculture

Contents

INTRODUCTION

Organic aquaculture is a sustainable farming practice that raises fish and seafood in environmentally friendly ways, adhering to strict organic standards. It focuses on maintaining the health of ecosystems and using natural materials and methods to avoid harming the environment. This approach promotes biodiversity, reduces pollution, and ensures the welfare of aquatic life. Organic aquaculture products are certified to guarantee they meet specific organic and environmental criteria (Sethi et al., 2023). This approach to fish farming embraces a comprehensive and environmentally friendly strategy, focusing on using natural methods for raising fish in both freshwater and saltwater settings. It prioritizes the long-term health and sustainability of the surrounding ecosystems, ensuring they remain viable for future generations. The techniques employed for breeding, treating, and caring for different fish species are eco-friendly and do not pose a threat to the environment. Over time, this form of aquaculture has gained popularity due to increasing concerns about the impact of farming practices on the natural world. Organic aquaculture products, recognized for their environmentally conscious production, are now widely available in markets around the world, experiencing a steady growth in consumer demand. This farming method is particularly prevalent in marine settings and various countries, such as Germany, Switzerland, and the United Kingdom (Sethi et al., 2023).

However, a significant issue arises with consumer perception, as there is confusion and skepticism regarding the labeling and standards of organic aquaculture products. To address these concerns, governments and regulatory bodies have developed specific labeling requirements. A designated certification body is responsible for the issuance of organic product seals, which are awarded only after a comprehensive review of the production process and verification that all standards and requirements have been met. This certification process involves a thorough assessment of various factors, including the compatibility of soil-based systems with aquaculture practices. There is a notable conflict between soil-based farming and intensive aquaculture, particularly in terms of environmental impact (Jun, Hao, Xia, Yan-Hua, & Ting-You, 2013). Despite its environmental benefits, organic aquaculture faces several challenges. These include the complexities of certifying young fish as organic, the high costs associated with organic feed, difficulties in obtaining juvenile fish, labor-intensive farming practices, increased risk of disease, and the substantial costs involved in obtaining certification. Nonetheless, the demand for organically farmed aquatic products continues to rise globally. Given this growing interest, many experts believe that, with further research and development, organic aquaculture could become an essential strategy for sustainable fish farming, offering a viable solution to current and future environmental and food security challenges. (Michèle, 2019).

8.1. ORGANIC AQUACULTURE PRODUCTS

Different countries follow different procedures when certifying their organic products. This means that there are standards that have to be met. The differences may be attributed to the fact that most of these countries have not come to an agreement on the standardization process. The Global Trust is the major organization tasked with the certification of various organizations. They usually develop the assessment procedures and certifications that need to be done so as to ensure that all the products generated have met international standards. (L. & S., 2018) Usually, information given by the Global Trust is readily available, so any person can make a personal inquiry.

The certificate issued is very important because it highlights a number of very important issues. It lists the type of treatments given to fish, whether the fish was given some antibiotics, whether proper disposal methods were used in eliminating feces from fish, the kind of feeding materials given to fish, the habitat in which fish were reared, the manner in which fish were raised, and the kind of handling practices used when rearing the fish. (L. & S., 2018) There is a variation in the rules followed by the different countries with regard to certification.

Though it may be beneficial for producers in various countries, it still presents a challenge to consumers who purchase these products from other countries, as they may not meet the various standards put in place. This explains why some companies have faced backlash from some of their consumers. When developing the guidelines with regard to certification, there are a number of things taken into consideration, including the species of animal in question. There are various species of fish that can be reared in aquaculture ponds. They include aquatic plants, finfish, mollusks, saltwater fish, shellfish, and freshwater fish.

Regulation of fish production is very different, as they cannot all be categorized as one. This is because there is a difference in the screening protocol followed when looking for pollutants in an aquatic medium. There is also a difference in the food supplies and the manner in which individual species of fish are monitored within the pond. The variation increased the need for bodies to develop different certification methods in matters of aquaculture. (Kaiser, 2012)

8.2. CHALLENGES IN ORGANIC AQUACULTURE

Organic aquaculture faces multiple challenges, including issues related to licensing and high production costs. Licensing controversies arise mainly because some seafood companies disagree with the strict regulations, particularly those that exclude wild-caught fish from being labeled as organic. (Kaiser, 2012) The contention stems from the fact that wild species may have been exposed to pesticides and chemicals in their natural environments, making it difficult to certify them as organic. In response to these complexities, the National Organic Standards Board decided in October 2001 that aquatic animals must be raised in clean environments and fed organic diets with minimal chemical exposure to qualify as organic. (Claude, Aaron, Jason, & Howard, 2005) This decision underscores that the entire ecological and production processes must adhere to organic principles for aquatic animals to earn organic certification. Farmed fish, even if raised under controlled conditions, face scrutiny regarding their feed, which must also be organic and free from chemical contaminants to be considered for organic labeling. (Lee, 2008)

8.3. CONSISTENT STANDARDS

By 2010, the organic aquaculture industry had developed around 80 different standards that were being applied across approximately 18 countries within the European Union. These standards, while aimed at promoting sustainable and environmentally friendly aquaculture practices, varied significantly from one country to another, reflecting the diverse approaches to organic farming and seafood production across the region. The goal of these standards is to ensure consistency and quality in organic aquaculture products, achieved through rigorous compliance and thorough inspections by relevant authorities. (Bush, 2013).

Over the years, these standards have undergone updates and revisions to reflect the evolving understanding of what constitutes sustainable aquaculture practices. In 2010, the European Union introduced new rules that aimed to harmonize and refine the criteria for organic aquaculture across its member states. These updated rules were designed to standardize practices, ensuring a more consistent application of organic principles in aquaculture operations. Also, various countries have established their own boards and regulatory bodies to oversee the implementation of these standards, further ensuring that organic aquaculture products meet the highest levels of environmental and ethical production. (Claude et al., 2020) (Figure 8.1).

For instance, in Canada, the board in charge of organic aquaculture is the Canada General Standards Board. The board is charged with the task of updating the various standards that should be maintained by producers. Some of the standards they generated in 2010 were rejected as those standards would allow the use of chemicals and antibiotics in the rearing of fish. The standards also allowed the use of nonorganic feeds in rearing these fish. The move was opposed as it would have severe effects on the environment. This would mean that net pen systems would be classified as organic (Pita, C. 2011). This was not the case in other countries, such as Denmark, where they placed very strict standards that would not allow the development of organic trout. As the feeds have to be organic, there may be a challenge in supplying fully organic feeds. For this reason, other alternatives have to be developed to meet the growing demand (Zanoli, 2019).

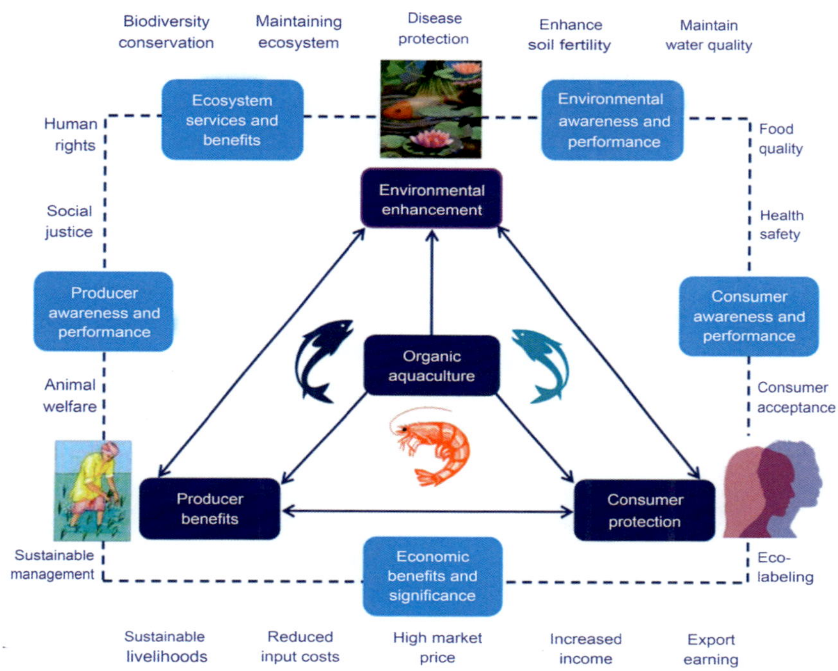

Figure 8.1. *Organic aquaculture productivity, environmental sustainability, and food security.*

8.4. ALTERNATIVES TO NON-ORGANIC FEED AND WASTE REMOVAL

The continuous supply of fully organic feeds may be a challenge. For this reason, there is a need to develop some alternatives that will ensure the continuous survival of aquaculture. This has been a major issue in aquaculture production, as it is not easy to find practical and sustainable methods. (Valsa & Namasivayam, 2009) This not only applies to food products; it also applies to the kind of treatment and spat used. It also looks into the matter of waste disposal.

Some of the potential veterinary alternatives used include the production cycle, limited allopathic treatments, chemical treatments, and homeopathic treatments. Currently, there are requirements that require farmers to reduce the use of unstainable fishmeal. (Costa-Pierce, 2008) This would mean that farmers will have to fully adopt the use of organic vegetables and other fish-based products. Researchers have been able to establish that organic feeds used for fish, such as salmon, provide a lot of benefits to the environment as they positively affect the fish's life cycle. The difficulty arises from the fact that the majority of these feeds are frequently not readily available. This may lead to the death of large numbers of fish. On a positive note, some of these feeds have been safely replaced. Organic aquaculture generally requires that both the feed and the fish are developed using organic means. (Valsa & Namasivayam, 2009).

More studies are being done to reduce the amount of non-sustainable feed that is available in the markets. Most of these feeds are replaced with organic vegetable proteins. This will see to it that there is a wide market for organic feeds. The availability of alternatives in the various markets will also help boost feed production. This has also enabled farmers to look into the option of using multi-integrated systems that may include the growing of plants through the use of aquaponics. This may take the form of farmers creating a shellfish bed ne□t to a finfish farm. This will allow better waste disposal as the shellfish will be able to benefit from the nutrients gotten. For a healthy ecosystem to be maintained, it is important that farmers look into the proper disposal of feed waste. For this reason, farmers regularly check the amount of feed waste found in the water. Water samples may have to be subjected to a series of tests at a frequency of about 3 to 4 months. The other challenge faced by farmers is the threat posed by wild fish that may find their way into these environments or some of the organic fish escaping from the ponds (Pita, 2011).

8.5. STATE OF DEVELOPMENT FOR ORGANIC AQUACULTURE

Organic aquaculture is considered one of the fastest-growing sectors in food production. Currently, it is estimated that this sector grows at a rate of 9%, which is higher when compared to the growth of terrestrially farmed meat and that of captured fisheries, which is at about 1.3%. Despite the high growth rate, there is still a challenge in the fact that the aquaculture sector has been seen to lag behind in terms of the diversity of fish reared and the amount or quantity of fish produced. The lag has been linked to the absence of a general or universal standard and accreditation criteria that will be used in general aquaculture production. It is a lot linked to the fact that most of the certifying bodies tend to be restricted, as there are several organizations that have been developed within those countries, namely Oceania, North America, and Europe. Most of these countries have been said to produce about 10% of the total quantity of fish produced. There is a challenge in getting organic aquaculture products that have been certified. (Rozana, Jan-Jan, Rusmani, & Nur, 2023).

Aquaculture has lagged behind the agriculture sector in terms of the quantity and diversity of certified organic produce. In 2000, it was estimated that about 5 000 metric tons of certified aquaculture products were produced in European countries. This is about 0.01 percent of the total aquaculture production. The total volume of organic aquaculture products in Europe in the year 2000 was in the range of 4,400 and 4,700 tons. The breakdown includes 100 tons of blue mussels, about 200–400 tons of carp, mostly produced in fresh water and mostly sold in domestic markets, about 100–200 tons of trout produced in German and Scottish markets, and the largest being the production of salmon, which was at about 4000 tons (P., 2007).

8.6. INTERNATIONAL ORGANIC AQUACULTURE STANDARDS

The body called IFOAM covers international standards for organic aquaculture by looking into its production. Most of the international standards were developed in the year 2000. Most of them have been adopted, and quite a number have not yet been adopted. The international aquaculture standards have seen to it that about 200 tons of shrimp are exported to other countries, as most of these products get their certification for the United Kingdom under the Ecuadorian farm. (Rossita, n.d.) Under the production process, there are experimental batches that are categorized as non-certified organic seabass. Most of these samples are not exported but are sold to domestic markets. On most farms, about 500–800 tons of salmon are produced and sold in other markets. Most of the information is generated from data inside Europe. There is little data on countries outside of Europe. With regards to aquaculture production, other than Europe, New Zealand is another major producer of such products. Among the aquaculture products produced in New Zealand, salmon was the first to be certified and is estimated to export about 500–800 tons of organic salmon. (Peter, 2008) The major market for New Zealand is Europe. However, over the years, there has been a notable decline in production. This has led to several non-European countries starting to develop their own organic aquaculture systems that meet both public and private standards. Most of these countries produce shrimp. They include Vietnam, Peru, Thailand, Ecuador, and Indonesia, among others. Some of them produce salmonids, as is the case in

Chile and Canada, while New Zealand produces mussels. There are a number of organizations that have been employed to carry out the certification of most of these products in compliance with organic aquaculture standards. With the development of such international standards, it is estimated that about 5000 metric tons of certified aquaculture products were produced in the year 2000, with Europe being a major producer (Pita, 2011) (Figure 8.2).

Figure 8.2. *Maroš Šefcovic, Executive Vice President of the European Commission in charge of the European Green Deal, Interinstitutional Relations, and Foresight, receives representatives of IFOAM (International Federation of Organic Agriculture Movements). Left to right: Eric Gall, Eduardo Cuoco, Jan Plagge, Silvia Schmidt, and Maroš Šefcovic.*

However, the absence of internationally recognized regulations and standards has contributed to the slow growth of organic aquaculture products. This has brought about deviations with regard to the production and handling of organic aquaculture production systems. (Jansson, 2010) This can be noted in the presence of both the European Union and the WHO/FAO Codex Alimentarius Commission, which is known to deal with setting the guidelines and standards for organic foods but has not tackled the issue of organic aquaculture. As this issue had to be addressed, the drafting of standards for organic aquaculture production was done by the International Federation of Organic Aquaculture Movements. The guidelines were first laid out in 1998. It was then accepted as a draft standard by IFOAM in the year 2000. These standards have been subjected to a series of revisions formed on the basis of deliberations, and some of them were fully voted in (Zanoli, 2019)(Jansson, 2010).

8.7. IFOAM GENERAL PRINCIPLES ON ORGANIC AQUACULTURE PRODUCTION

There are various principles that govern organic aquaculture production, which are as follows:

8.7.1. Conversion to Organic Aquaculture

Within the draft, converting an aquaculture system to organic involves the development of farming practices that will foster and maintain a viable and sustainable aquatic ecosystem. The draft also defines the term conversion period, which is the period between the time the organic management process is started and the certification of the production process. There are various production methods used in aquaculture that tend to vary depending on the organism's biology, the geographical conditions, the technology used during the process, the time span, and the ownership structure. All of these aspects influence the time during which the conversion process takes place (Pita, 2011).

8.7.2. Basic Conditions

During the production process, there are management practices put in place. The organism's ethological and physiological needs are what govern the majority of these practices. (Jansson, 2010) The ecosystem should be such that the organisms are able to satisfy their basic behavioral needs. The adopted practices will influence the level of production and the rate of growth within which the animal will be maintained, while also ensuring the well being and good health of these organisms. With the management of such systems, the farmer needs to take care so that the introduction of non-native species does not affect the natural ecosystems. (James, Helen, & James, 2006).

8.7.3. Location of Production Units

For an animal to be termed organic, there is a great need for the environment to be organic. This is very necessary, as the health of the aquatic environment is to be fully organic, and the same applies to the terrestrial and surrounding ecosystems. (Paulo et al., 2019).

8.7.4. Location of Collecting Areas

With the production of organic products, there is a great need for the surroundings to be well maintained. This means that the waste should be properly managed.

In most cases, there are open collecting areas that have been certified by the government as unpolluted, sustainable, and stable environments. (Stephanie & Paul, 2009).

8.7.5. Welfare and Health

During the production of organic products, management practices should be maintained at a higher level so that the fish species have a higher resistance to diseases and can overcome any infections that can occur. This means that the techniques used during the management process will affect the speed and level of production, which will generally affect the health and welfare of such organisms. In most cases, aquatic organisms require little handling. This means that the mode of treatment in case of injury or disease will determine the well-being of the animal (Pita, 2011).

8.7.6. Breeds and Breeding

The procedures and strategies used in organic aquaculture should ensure that normal practices are maintained. The same applies to the breeding process. For this reason, the normal breeding methods are encouraged. (Xianwen & Kyrre, 2018).

8.7.7. Nutrition

When feeding aquatic animals, it is important to ensure that they get feed that will meet all their nutritional requirements. This means that they should be given a balanced diet, and the feeds should be of good quality. (b., n.d.) The feeds should be given in such a manner that the natural feeding behaviors are well maintained while ensuring that there are minimal losses to the environment. Feeds given to such organisms include organic products of organic food processing. They also include wild marine feeds from various resources that can be consumed by humans.

8.7.8. Harvesting

The method followed when harvesting these aquatic animals will determine whether or not they will be categorized as aquatic or not. The harvesting process should be done in such a way that minimum stress is placed on the organisms while in the ecosystem. This means that the natural balance of these animals is well maintained (Zanoli, 2019).

8.7.9. Transportation of Living Marine Animals

The management of organic aquaculture products also takes into consideration the transportation process. This means that the method of transportation used should be the most suitable and should vary depending on the species of animal in question. The transportation process includes the type of water used. The water used should have the right amounts of oxygen, salinity, and temperature. It is also necessary to ensure that the frequency, duration, and distance are well minimized (David et al., 2018).

8.7.10. Slaughter

Slaughter is the final stage of the production process. When conducting the slaughtering process, the farmer should ensure that the activities are done in such a way that the organism experiences minimum suffering and stress. The process is done following a number of techniques and management practices that are done using various considerations, such as the etiology and physiology of the organism in question (David et al., 2018).

The adoption of all these processes is taking place at a gradual, though very slow, rate. About ten years after the promulgation of the States Organic Food Production Act, the United States Department of Agriculture published the National Organic Program, which lists the final rule on the National Standards for Organic Livestock and Crop Production. The challenge with these laws is that they do not take into consideration aquatic animals, as they do not cover them. It is anticipated that some amendments will have been made to allow better aquaculture practices. This later led to the National Organic Standards Board establishing the Aquatic Task Force and Aquaculture Working Group. The task force would have seen the various key issues that would later allow various recommendations to be made to allow better aquaculture production processes. This saw the submission of the NOP being done in 2001. (Robert & Robert, 2021) The recommendations saw an allowance being set that will govern the kind of fish meal and oils that are incorporated in the diets of such animals and is considered a constraint for producers of such products living in the United States. This led to the development of national standards in the United States that will mandate the handling of livestock within the area of organic aquaculture, such that these animals will be given feeds that are 100% organic. As the law applies to livestock, it also affects aquatic animals. This means that both the fish oil and fish meal should be organic in nature (Robert & Robert, 2021).

A lot of efforts have been made by the Aquatic Task Force and working group. This can be seen in the laws that have been made. The challenge is that there is a sense of lack of clarity with regard to the drafts and proposals made by the NOP with regard to organic aquaculture. There are certain acts that deal with the handling of food. For instance, there is an Organic Food Production Act stating that a person is allowed to label or sell agricultural products as organically produced on condition that the final product is managed and handled with reverence for organic principles. This gives a clear guideline for farmers who want to transition from normal agricultural practices to organic production. It also helps them fully acquire and earn organic certification. It was quite a process before the USDA came to a decision on granting producers the seal that shows their products are organic in nature. (Paul & David, 2008).

As most countries were not able to come to a consensus with regards to these standards such that they have universally agreed on the standards they would follow, there has been a challenge with the certification of these products. This has left the individual member states as well as the private governmental certifying agents to develop their own rules with regard to the standards that would be used to govern the certification of other products. This e□plains why there are several organizations that certify farmers as well as accreditation bodies. It is worth noting that the major challenge faced by most of the associations is the fact that several drafts are made, but they take a long time before they are fully adopted, as is the case with the Soil Association, which generated its first organic awuacu6 standard that took time before it was implemented. The laws made would affect the labeling and certification of trout and organic salmon before they reach the markets. The developed salmon aquaculture standards that are used, along with those of the United Kingdom, where private certifiers are used, have generated more recognition across the world. Both Australia and France have been able to generate national aquaculture standards (Zanoli, 2019) (Figure 8.3).

Figure 8.3. *Salmon is an example of the most popular species of aquatic animals sold.*

Naturland is the pioneering private certifying body that is based in Germany and is involved in the development and promotion of organic aquaculture in Europe. The association started its activities in 1995, which saw the development of standards that will be used in organic aquaculture and the initial certification of various products. It is also involved in trench production, which is also carried out in Germany through the use of some traditional techniques. Ireland has also developed various laws and standards that govern the production of salmon and mussels as well. This has later seen the growth of markets for these fish products in various parts of the world. The association based in Germany later extended its activities so that they could be accessed by developing countries, which led to the development of standards that will be used in shrimp production in ponds (Paul & David, 2008).

Among the aquaculture products sold in large markets, shrimp is among them. Shrimp products are usually sold in international markets as an aquaculture commodity. They currently hold a good market price in that, when they are properly managed and harvested, they can be sold at a good price. Developing countries contribute a large percentage of the total number of shrimp produced globally, with the major producers being Latin America and Asia. Production is boosted through support from other countries, such as Germany, through the Naturland Base Association. The organization is said to be piloting its first production of organic shrimp. This opens doors for other countries to be part of the global shrimp production. Some of these countries include Viet Nam, Peru, and Indonesia. In May 2000, Ecuador was able to obtain certification to produce and sell organic shrimp, with its largest market being the United Kingdom. There is a growing increase in the organic aquaculture sector in terms of size, with the expansion extending to the private sector. This has allowed the generation of both private and non-private certifying bodies whose standards are different depending on their requirements and the type of species being reared. In Ireland, there is a fisheries advisory commission that works on organic fish farms (Pita, 2011).

There are many moves made by various governments to see the development of the aquaculture sector. This includes the establishment of an ad hoc committee that will work on the organic fish farms. This was to be done by a working party that had several roles to play, among them being the complication of various material on the status and development of organic fish farms, consideration of the various market conditions entailing organic fish farms, reviewing and discussing the various technical aspects of organic fish production, and also laying out the various research needs. The party will also take into consideration the technical aspects of the various guidelines and will take on the role of drafting the guidelines that will oversee the production of organic fish. As of the year 2000, the largest production of organic aquaculture was from European countries, which was about 2 million tons. There are concerns that have been raised about the need for the EIFAC to look into organic fish farms so as to boost growth rates that were at that time at 25 %. This would see the farming of various species of fish that include coregonids, pikeperch, pike, tench, and carp. Most of these species would be reared in various countries, including the United Kingdom, Ireland, and Germany. Most organic aquatic farms have adopted the use of culture systems, which are, in essence, traditional ponds used in rearing carp and other fish products reared in offshore cages or earthen ponds. Some of the achievements seen in the working part include the dissemination of questionnaires to the EIFAC with the aim of obtaining reviews of organic fish farming done in European countries. In the original copy, the questionnaire only covered freshwater environments and later developed to include marine and brackish environments. This led to a number of workshops being done, including the workshop in Nürnberg. The workshop held in February 2000 was very essential in highlighting the various issues faced in organic farming. Some of the issues discussed include the national aspects, which covered the use of acceptable levels of fish feeds, the utilization of natural antio□idants, the use of alternative protein sources, and various environmental factors, etc. It also looked into the matter of holding facilities, which looked into the type of materials used in construction, the role played by oxygen in the oxygenation process, stocking densities, and water quality aspects. Under matters post-harvesting, the part looked into the slaughter process, the handling of fish products, the transportation process, and the processing of these substances. The party also seeks to facilitate the scientific scrutiny of various themes. This has highlighted the difference between the farmer, certifier, and all relevant stakeholders and outlined their various roles by interpreting the field of organic aquaculture. This led to the great need for farmers to adopt the various basic principles as well as the production standards (Pita, 2011; (Paul & David, 2008).

8.8. ORGANIC AQUACULTURE AND THE ENVIRONMENT

The basis of organic aquaculture is to make sure that aquatic products are produced through natural means while ensuring minimal disturbances to the environment. The kind of environment in which the various species of aquatic animals are reared will influence whether or not it is organic. The switch from ordinary conditions to organic aquaculture may tamper with the environment if not done properly. This is why there are laid out procedures to be followed with the development of such systems. The farmer may need to conduct more research on the best strategies for adopting organic systems. The kind of feed used and the method used in the feeding process will influence the environment either directly or indirectly. For instance, when feeds are given to the various species in certain ponds, they should not be allowed to last there for longer durations, as it would tamper with the natural ecosystems. There are specified durations for which they should stay in the water. If the feeds are left in the water for too long, it can lead to overfeeding and pollution of the ecosystem (Paul & David, 2008) (Figure 8.4).

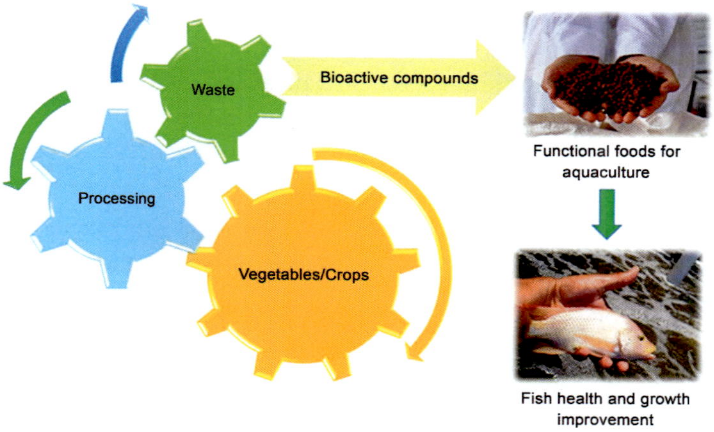

Figure 8.4. *Organic aquaculture helps maintain a clean environment.*

There are various ecosystems involved in the rearing of aquatic animals. They include fresh water and salty water. Some farmers opt to use brackish environments. Marine and brackish environments contribute a large percentage of the total aquatic environments distributed across the world. Some of the aquatic products derived from marine environments are mollusks, aquatic plants, crabs, shrimp, and salmonids. Of the total number of such products produced, some percentage is said to be certified organic aquaculture products, mostly generated in Europe. Most of the certified aquatic products are generated in brackish and marine waters. Research shows that a large percentage of both fresh and marine water species have not yet been captured. This means that there could be several other species of aquatic animals that have not yet been discovered. This calls for more research to be done. This will mean that both the fresh and salty water sources will be conserved and can be used by future generations for human consumption and general agriculture production that entails the rearing of both livestock and aquatic animals (Pita, C. 2011). When it comes to marine production, there are more species of animals that can be discovered than mollusks and organics. The newly harvested species can be used for direct human consumption or it can be used to feed other animals within animal husbandry (Zanoli, 2019).

Like the livestock sector, for the organic aquaculture sector to successfully coexist with other food production sectors, more efforts have to be made to allow the generation of organic feeds from aquatic species. These feeds have to be packed with the right kind of nutrients and have the most suitable nutritional profile. This means that if aquatic animals feed on meat, then the organic feed has to incorporate meat, as is the case with trout and salmon. Some organic feeds adopt the use of fish meal and fish oil. With the generation of animal feeds used in organic production, there were some concerns raised. They include whether feeds will be certified after being caught in the wild, the use of certified organic feeds, the acceptable levels of fish meal and fish oil that can be used, the long-term suitability and ethics of the production of organic carnivorous fish species, and the transfer of lipid and essential protein sources across the world (Egidijus, Jannicke, Roger, & Anne-Kristin, 2020).

The major challenge with aquatic production is pollution and overfishing. All of these activities have led to a reduction in the total number of fish found in the sea. This has tampered with the supply of fish and reduced the total number of sea fruits that can be sold in various markets. The demand for such products has to be met. This means that an expansion should be experienced in the aquaculture sector. Studies show that intensive aquaculture contributes to the largest percentage of sea pollution, and some of the challenges being experienced include the devastation of various mangrove forests so as to construct ponds. This has led to many farmers and fishing communities being damaged. Poor management of such farms also contributes to the spread of diseases in the aquatic farms and eventually the abandonment of these ponds, as they will be destroyed through the uncontrollable accumulation of pesticides and nutrients. Another cause of pollution is the unrestrained use of antibiotics, which has led to resistance against pathogens building up. This applies to both the farmed organisms and those caught in the wild (Pita, 2011).

Among the major aquatic products sold across the world are shrimp. They are usually produced in Latin America and Southeast Asia. The major producers of such products are Ecuador and Thailand. They export large numbers of such products, such that the money generated has been used to sustain their ecosystems. This has led to the rapid growth of aquacultural production, though it has been linked to certain ecological and social-economic effects. With consumers becoming more aware and sensitive about what they eat, threats (both social-economic and environmental) posed by intensive aquaculture activities have found their way to various environmental bodies and other NGOs, etc. This has seen some companies boycott the use of various aquacultural products. This has led to shrimp farming becoming a contentious matter. Due to this fact, the organic community is also affected, and therefore some farmers may find it hard to transition from traditional methods to organic systems. It may also affect the likelihood of these farming methods generating positive results. This has led to various efforts being put in place to ensure that shrimp production does not affect general aquaculture production. This led to the generation of organic standards that will govern shrimp production. Later on, IFOAM formulated drafts of standards that will be used in organic fish farming. The first shrimp project was done in 2000, which saw their certification by Naturland (Pita, 2011). When farmers are able to fully adopt organic principles in the aquaculture process, then the generated products will be able to sustain and manage the fisheries as per the international management guidelines. They will be able to get these products from locally available markets, though they are said to not be suitable for human consumption, and they will be free from unwanted contaminants and synthetic additives. The production process usually makes use of public water bodies, especially in the rearing of aquaculture products in cages and nets. This kind of farming method is very sensitive as they are vulnerable while in the open, thereby presenting a challenge as many of these organisms tend to escape from such systems. There is also the challenge of the generation and release of effluents. There is also the challenge of the oceans being e□posed to chemical drift. These challenges have not been considered barriers but as some of the limitations faced. This is a major issue in the United States. These species should be allowed to behave naturally so that they can grow to full term and grow in large numbers (Zanoli, 2019).

8.9. LONG-TERM PROSPECTS FOR CERTIFIED ORGANIC AQUACULTURE PRODUCTS

With the growing demand for purely organic products and feeds, it is anticipated that more people will venture into organic aquaculture. This means that more investments will be made toward boosting the growth of this sector. This means that more research will be conducted to look into the various ways in which the various farming practices can be implemented so as to boost the sector. This also means that more advanced methods can be developed to enable effective organic aquaculture production. It also provides an avenue for one to look into the possibility of adopting technology in the rearing process, either in feeding the aquatic animals or in cleaning the tanks used in the process. This may boost the compound annual growth that will see the sale of a larger number of species of aquatic animals. Governments are looking into dealing with the problem of certification. This may be dealt with by adopting more organizations that will offer such certificates to farmers. With more elaborate standards being laid out, it becomes easier for a farmer to certify their products at a higher rate. All of this will, in the long run, allow farmers to certify their products at lower prices, leading to the availability of several certified products on the market. This may see organic aquaculture production increase by about fourfold, resulting in the production of over 194 million tons of such products by the year 2030. Most of the produce is from developed countries. It is also expected that markets for such products will grow mostly in North America and Europe. It is also expected that such markets will grow in countries such as Singapore, Japan,

New Zealand, and Australia. It is anticipated that the growth will be pushed and fueled by the growing awareness of environmental pollution and more people becoming health conscious (Pita, 2011).

Certification of such products goes a long way in proving their safety. These products will be tested and proven to be organic in nature. However, there is a great need for governments to ensure that there is no form of corruption that will see defective or contaminated products being termed organic and safe. This may affect the markets of the countries from which these products come. However, the drastic growth of such markets may be noted because certification was fully and quickly adopted by the developing countries, and there is a method of ensuring that these products are produced through earnest methods. Developed countries, therefore, need to move their certified products from the experimental stage to fully enter markets. As for developing countries such as China, it is worth noting that the projected growth rate of production is relatively high, as per studies done within her 3000-year history. The country is quickly adopting systems that favor the rearing of both fresh and salty aquatic animals on a large scale while ensuring the products are fully organic. China is also known to have made use of farming strategies that date back several years (Zanoli, 2019).

8.10. ADVANTAGES OF ORGANIC AQUACULTURE PRODUCTION

There are several advantages to organic aquaculture production, one of which is that it preserves the environment. One of the principles of organic aquaculture is that the environment should not be tampered with. It places much emphasis on the need to ensure that the natural behavior of animals in the ecosystem is not affected. This is beneficial in ensuring that the environment is well maintained, as the products used in the rearing process are fully organic, as are the methods used. Maintaining the natural ecosystems ensures that these animals and ecosystems will be able to survive and be used by future generations. It also ensures the good health of the organism, as a healthy environment will ensure that these animals are not affected by diseases or pests. This makes the farming process very holistic (Pita, 2011).

Another advantage of organic aquaculture is that it can be used to boost food security. Currently, there is a big issue of food insecurity. By boosting the organic aquaculture sector, a consistent supply of such products is ensured. These products will be used to add to the general amount of feed available for all. If developed countries were able to fully adopt this method of farming, then the issue of food insecurity would be a problem of the past. Organic aquaculture is also useful in boosting income for farmers and the state in general. Farmers sell their products in both local and international markets. By exporting their produce to countries abroad, the country gains revenue that can be used to drive the economy, as noted in Europe. Consumers also benefit from the fact that these products are very healthy and are free from any form of contamination (Pita, 2011).

REFERENCES

1. Bush, R. (2013). Certify sustainable aquaculture? *Science, 341*, 1067–1068. Retrieved from: https://www.science.org/doi/abs/10.1126/science.1237314 (accessed on 3 August 2024).
2. Claude, A., Jason, & Howard, A. (2005). Certification issues for some common aquaculture species. *Taylor & Francis Online, 13*(4), 231–279. Retrieved from: https://www.tandfonline.com/doi/abs/10.1080/10641260500326867 (accessed on 3 August 2024).
3. Claude, L., Brent, D., David, L., George, M., & n. (2020). Achieving sustainable aquaculture: Historical and current perspectives and future needs and challenges. *Wiley Online Library, 51*(3), 578–633. Retrieved from: https://onlinelibrary.wiley.com/doi/abs/10.1111/jwas.12714 (accessed on 3 August 2024).
4. Costa-Pierce, B. (2008). An ecosystem approach to marine aquaculture: A global review. *Rem Main*. Retrieved from: http://rem-main.rem.sfu.ca/papers/knowler/i0339e.pdf#page=89 (accessed on 3 August 2024).
5. David, J., Wenbo, R., Abdullah, N., & Francis, G. (2018). Sustainable intensification of aquaculture value chains between Asia and Europe: A framework for understanding impacts and challenges. *ScienceDirect, 493*, 338–354. Retrieved from: https://www.sciencedirect.com/science/article/pii/S0044848617325292 (accessed on 3 August 2024).
6. Egidijus, J., Jannicke, R., & Anne-Kristin, B. (2020). Fish and fish waste-based fertilizers in organic farming–With status in Norway: A review. *ScienceDirect, 115,* 95–112. Retrieved from: https://www.sciencedirect.com/science/article/pii/S0956053X20303913 (accessed on 3 August 2024).
7. James, H., & James, A. (2006). Seafood certification and ecolabelling: A new wrapping on fisheries resource management? *Oregon State University Library*. Retrieved from: https://ir.library.oregonstate.edu/concern/conference_proceedings_or_journals/br86b464m (accessed on 3 August 2024).
8. Jansson, T. (2010). Green feed in the marine fish farming. *Epsilon*. Retrieved from: https://stud.epsilon.slu.se/1020/ (accessed on 3 August 2024).
9. Jun, H., Hao, X., Xia, Y. H., & Ting-You, T. (2013). Organic aquaculture in China: A review from a global perspective. *ScienceDirect, 414–415*, 243–253. Retrieved from: https://www.sciencedirect.com/science/article/pii/S0044848613004171 (accessed on 3 August 2024).

10. Kaiser, M. (2012). The ethics and sustainability of aquaculture. In *The Philosophy of Food*. Retrieved from: https://books.google.com/books?hl=en&lr=&id=FjreD88VN74C&oi=fnd&pg=PA233&dq=Organic+aquaculture+is+a+sustainable+farming+practice+that+raises+fish+and+seafood+in+environmentally+friendly+ways,+adhering+to+strict+organic+standards.&ots=MZ9HD991ba&sig=HV222KEXOz5yBihlR4mubSS0nkA (accessed on 3 August 2024).
11. L., & S., A. (2018). Organic aquaculture as promising trend of the fish industry development. *Agrobiology, 53*(2), 326–336. Retrieved from: http://www.agrobiology.ru/articles/2-2018lagugtkina.pdf (accessed on 3 August 2024).
12. Lee, M. (2008). Aquaculture certification. *Oxford University Press*. Retrieved from: https://books.google.com/books?hl=en&lr=&id=wGh□zFX3rG4C&oi=fnd&pg=PA106&dq=Organic+aquaculture+is+a+sustainable+farming+practice+that+raises+fish+and+seafood+in+environmentally+friendly+ways,+adhering+to+strict+organic+standards.&ots=LzN7Hd1TMi&sig=ALvwQpgV7IBWAYh5VTNaA_vTTNI (accessed on 3 August 2024).
13. Michèle, S. (2019). Recirculation aquaculture systems: Sustainable innovations in organic food production? *SpringerLink, 4*, 67–84. Retrieved from: https://link.springer.com/article/10.1007/s41055-019-00054-4 (accessed on 3 August 2024).
14. P., H. (2007). Feeding farmed salmon: Is organic better? *ScienceDirect, 272*(1–4), 399–416. Retrieved from: https://www.sciencedirect.com/science/article/pii/S0044848607005169 (accessed on 3 August 2024).
15. Paul, G., & David, P. (2008). Challenging the aquaculture industry on sustainability. *Planet4*. Retrieved from: https://planet4-canada-stateless.storage.googleapis.com/2018/06/challenging-aquaculture.pdf (accessed on 3 August 2024).
16. Paulo, N., Nohreen, J., Jeanette, R., Rachelle, P., & Philippines, S. (2019). Good Aquaculture Practices Adoption and Certification of Shrimp Aquaculture Farms in Bulacan, Philippines: Status, Issues and Prospects. *SSRN, 15*(2), 11–36. Retrieved from: https://papers.ssrn.com/sol3/papers.cfm?abstract_id=3539246 (accessed on 3 August 2024).
17. Peter, D. (2008). Life cycle considerations for improving sustainability assessments in seafood awareness campaigns. *SpringerLink, 42*, 918–931. Retrieved from: https://link.springer.com/article/10.1007/s00267-008-9148-9 (accessed on 3 August 2024).
18. Robert, L., & Robert, S. (2021). Sustainable aquaculture.

SpringerLink. Retrieved from: https://link.springer.com/chapter/10.1007/978-3-030-84216-1_5 (accessed on 3Auggust 2024).

19. Rossita, A. (n.d.). Aquaculture ecosystems: Adaptability and sustainability. *Wiley Online Library*. Retrieved from: https://onlinelibrary.wiley.com/doi/pdf/10.1002/9781118778531 (accessed on 3 August 2024).
20. Rozana, J. J., Rusmani, R., & Nur, S. (2023). Consumers' preference and willingness-to-pay for GAqP-compliant farmed fish produce: Evidence from Malaysia. *ScienceDirect, 568*, 739305. Retrieved from: https://www.sciencedirect.com/science/article/pii/S0044848623000789 (accessed on 3 August 2024).
21. Sagar, P., Pijush, M., & Malay, K. (2023). Chapter-9 Sustainable aquaculture and health benefits through-organic fish farming. *ResearchGate*. Retrieved from: https://www.researchgate.net/profile/Pijush-Payra/publication/374144319_Sustainable_Aquaculture_and_Health_Benefits_through-Organic_Fish_Farming/links/65676edb3fa26f66f439b5df/Sustainable-Aquaculture-and-Health-Benefits-through-Organic-Fish-Farming.pdf (accessed on 3 August 2024).
22. Sethi, S. N., Meena, D. K., Sahoo, A. K., Nayak, A. P., Sahoo, S. N., Tiwari, P. K., Mohanta, K. N., Pillai, B. R., & Swain, S. K. (2023). Organic aquaculture: An overview. *ScienceDirect, 2023*, 561–598. Retrieved from: https://www.sciencedirect.com/science/article/pii/B9780323991452000094 (accessed on 3 August 2024).
23. Stephanie, A., & Paul, D. (2009). Animal welfare and organic aquaculture in open systems. *SpringerLink, 22*, 437–461. Retrieved from: https://link.springer.com/article/10.1007/s10806-009-9169-2 (accessed on 3 August 2024).
24. Valsa, G., & Namasivayam, K. (2009). Functional options for sustainable shrimp aquaculture in India. *Taylor & Francis Online, 17*(3), 336–347. Retrieved from: https://www.tandfonline.com/doi/abs/10.1080/10641260802715072 (accessed on 3 August 2024).
25. Xianwen, Z., & Kyrre, M. (2018). Labeling farmed seafood: A review. *Taylor & Francis Online, 22*(1), 1–26. Retrieved from: https://www.tandfonline.com/doi/abs/10.1080/13657305.2017.1356398.

INDEX

F

G

T

U

V

W

Y

Z

FSC
www.fsc.org
MIX
Papier | Fördert gute Waldnutzung
FSC® C083411

Kolibri 360 GmbH
Ferdinand-Jühlke-Straße 7
99095 Erfurt, Deutschland
produktsicherheit@kolibri360.de